Hf-BASED HIGH-*k* DIELECTRICS

Process Development, Performance

Characterization, and Reliability

© Springer Nature Switzerland AG 2022

Reprint of original edition © Morgan & Claypool 2005

Hf-Based High-k Dielectrics, Process Development, Performance Characterization, and Reliability

Young-Hee Kim and Jack C. Lee

ISBN: 978-3-031-01424-6 Kim/Lee, Hf-Based High-k Dielectrics (paperback)
ISBN: 978-3-031-02552-5 Kim/Lee, Hf-Based High-k Dielectrics (e-book)
Library of Congress Cataloging-in-Publication Data

First Edition
10 9 8 7 6 5 4 3 2 1

Hf-BASED HIGH-k DIELECTRICS
Process Development, Performance Characterization, and Reliability

Young-Hee Kim
IBM T. J. Watson Research Center,
Yorktown Heights, New York

Jack C. Lee
The University of Texas at Austin

ABSTRACT

In this work, the reliability of HfO_2 (hafnium oxide) with poly gate and dual metal gate electrode (Ru–Ta alloy, Ru) was investigated. Hard breakdown and soft breakdown, particularly the Weibull slopes, were studied under constant voltage stress. Dynamic stressing has also been used. It was found that the combination of trapping and detrapping contributed to the enhancement of the projected lifetime. The results from the polarity dependence studies showed that the substrate injection exhibited a shorter projected lifetime and worse soft breakdown behavior, compared to the gate injection. The origin of soft breakdown (first breakdown) was studied and the results suggested that the soft breakdown may be due to one layer breakdown in the bilayer structure (HfO_2/SiO_2: 4 nm/4 nm). Low Weibull slope was in part attributed to the lower barrier height of HfO_2 at the interface layer. Interface layer optimization was conducted in terms of mobility, swing, and short channel effect using deep submicron MOSFET devices.

KEYWORDS

Dynamic reliability, High-k dielectrics, HfO_2, Interface engineering, Mobility, MOSFET, Soft breakdown, TDDB, Weibull slope

Dedication to my wife, Mi-Sun, and children, Hannah, Hein.

Contents

Preface

Chip density and performance improvements have been driven by aggressive scaling of semiconductor devices. In both logic and memory applications, SiO_2 gate dielectrics have reached minimum thickness due to direct tunneling current and reliability concerns. Therefore high-k dielectrics have attracted a great deal of attention from industries as the replacement of conventional SiO_2 gate dielectrics. So far, many materials have been evaluated and Hf-based high-k dielectrics appear to be one of the promising materials for gate dielectrics. However, many issues were identified and more thorough researches were carried out on Hf-based high-k dielectrics. For instance, mobility degradation, charge trapping, crystallization, Fermi level pinning, interface engineering, and reliability are some of the critical issues. In this research, the reliability of HfO_2 was explored with poly gate and dual metal (Ru–Ta alloy, Ru) gate electrode. Hard breakdown and soft breakdown were compared and Weibull slope of soft breakdown was investigated under constant voltage stress. Dynamic stressing was used. It was found that the combination of trapping and detrapping contributed the enhancement of the projected lifetime. The results from the polarity-dependence studies showed that the substrate injection exhibited a shorter projected lifetime and worse soft breakdown behavior, compared to the gate injection. The origin of soft breakdown (first breakdown) was studied. The results suggested that the soft breakdown may be due to one layer breakdown in the bilayer structure (HfO_2/SiO_2: 4 nm/4 nm). Low Weibull slope was in part attributed to the lower barrier height of HfO_2 at the interface layer. Interface layer optimization was conducted in terms of mobility, swing, and short channel effect using deep submicron MOSFET devices. In fact, Hf-based high-k dielectrics could be scaled down to below EOT of $\sim$10 Å with excellent characteristics. However, it is still necessary to understand and separate the intrinsic properties from the extrinsic properties.

Acknowledgments

Dr. Kim would like to thank the advisor, Dr. Jack C. Lee and his colleagues Choong Ho Lee, Katsunori Onishi, Chang Seok Kang, and Hag-Ju Cho. All works in this publication have been accomplished with the help of accomplished with the help of colleagues, K. Katsunori, Chang Seok, Rino, Hag Ju, Akbar, Chang Yong, Se-jong, and Chang-Hwan as well as Veena Misra's group at NC State University.

CHAPTER 1

Introduction

1.1 FRONT-END DEVICE TECHNOLOGY EVOLUTIONS

For the past few decades semiconductor technology has been led by conventional scaling. Scaling, of course, has been aimed toward higher speed, lower power, and higher density. The continued scaling could be achieved through painstaking research and development effort and investment. However, as scaling approached its physical limits, it got more difficult and challenging [1–7]. In front end, the physical origins of these limits are primarily in the tunneling currents, which leak through the barrier in devices when it becomes very small, and in the thermally generated subthreshold current. Scaling of Vth also increases subthreshold current of devices and in turn leads to high-power consumption when the device is off. In such a small geometry, the number of dopants for a given area could possibly increase the standard variation of Vth resulting in large process variation [8–14]. Therefore, tremendous research has been carried out to investigate the alternatives, and this has led to the introduction of new materials and concepts to overcome the variations. In order to reduce gate leakage current and parasitic capacitance, high-k dielectrics and silicon on insulator were explored [15–35]. Also to increase inversion capacitance, drive current, and mobility, metal gate, double gate (FinFET), and strained Si were introduced (Fig. 1.1) [36–59]. In fact, conventional CMOS scaling is no longer driving force of technology evolution.

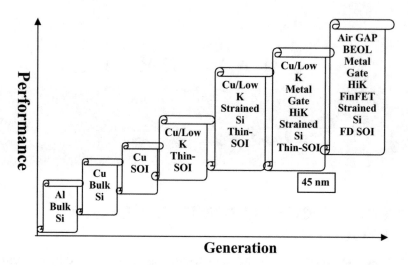

FIGURE 1.1: Projected technology evolution of IBM microelectronics from *Process Outlook Forum* April 15, 2003.

1.1.1   Substrate

Silicon on insulator (SOI) has shown lots of advantages as compared to bulk Si CMOS devices (Fig. 1.2). The reasons for performance improvements with SOI are mainly reduction of parasitic capacitance and easy isolation. Some of the recent applications of SOI include high-end microprocessor, low power, radiofrequency (RF) CMOS, embedded DRAM (eDRAM), and SiGe bipolar device. It is often compared like partially depleted SOI vs. fully depleted SOI with regards to manufacturability, design point (high Vth, dynamic Vth), breakdown voltage, short channel effect (SCE), kink effect, body contact, and history dependence. In recent

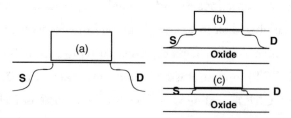

FIGURE 1.2: (a) Conventional bulk Si, (b) partially depleted SOI, and (c) fully depleted SOI.

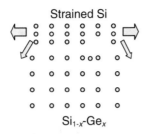

FIGURE 1.3: Biaxial tensile of strained Si on Si–Ge substrate.

research, ultra thin body SOI (6 nm) was demonstrated, which has good switching capability and SCE [60, 61]. Above all, SOI offers better power-delay product in comparison to bulk Si, very high resistivity substrates to achieve low loss passive elements, low power RF circuits, and so on.

Strained Si has been known to enhance carrier mobility in Si under biaxial tensile strain (Fig. 1.3) [40–59]. The theory is still evolving, and many processes have been introduced. It usually gets 30–70% mobility increase compared to that of a bulk CMOS device. It can be implemented with SOI for higher technology node. But there are many issues also like thermal budget control, defect control, biaxial vs. uniaxial strain, stress liner effect (SiN), and fabrication cost.

1.1.2   Gate Dielectrics

Since SiO_2 approached its physical limit, alternative dielectrics have been introduced, such as Al_2O_3, Ta_2O_5, ZrO_2, HfO_2, and their silicates, which meet stringent requirements including thermal stability, large band gap, and compatibility to conventional CMOS process. Moderate dielectric constant (k) materials (15–30) were preferred due to fringing field induced barrier lowering (FIBL). It was reported that there is a universal relation between k value and breakdown properties. In general, as k increases, barrier height and breakdown strength decreases (Fig. 1.4). It was explained that a material structure plays an important role of breakdown. In addition, it was shown that different charge fluence by different barrier height changes breakdown properties.

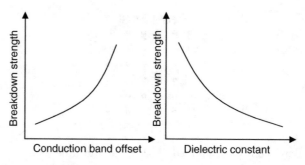

FIGURE 1.4: Universal relation between band offset and breakdown strength.

Single metal oxides (Al_2O_3, Ta_2O_5, ZrO_2, HfO_2) have been studied for decades. It was shown that crystallization usually happens at high temperature, except that of Al_2O_3. In addition, interface layer thickness due to subsequent thermal processing makes scaling problematic. However, by controlling and preventing interface layer growth, EOT (equivalent oxide thickness) of below 10 Å using HfO_2 was reported and proven to be scalable and compatible to modern CMOS transistor. So far, many process improvements have been reported including nitridation profiling, D_2/Forming gas anneal, and O_3 treatment at interface. The introduction of new processes has improved the performance, crystallization temperature,and interface state density.

Binary and ternary oxides (HfAlO, HfSiO, HfSiON, ZrON) have, in general, superior crystallization temperature compared to single metal oxides. However, these often exhibit unwanted effect such as phase separation at high temperatures. By incorporating nitrogen, these oxides can get over phase separation and stay in amorphous state at high temperature, while keeping barrier height high enough (Table 1.1 and 1.2). So far, HfSiON and HfO_2 with appropriate treatments have been successfully demonstrated to the choice of technology.

1.1.3   Gate Electrodes

Metal gate electrodes were investigated in order to eliminate poly depletion effect resulting in the reduction of inversion capacitance. In the case of single

TABLE 1.1: Role of N Profiling in HfO$_2$ [62]

MERITS	PROFILE	ISSUES
Impurity diffusion (B,P) ↓ Impurity diffusion ↓, Oxygen diffusion (interfacial oxide) ↓, Dielectric constant of interface ↑, Leakage current ↓, Breakdown voltage ↑, Improve scaling	Top	N out-diffusion during process
Impurity diffusion ↓, Oxygen diffusion (interfacial oxide) ↓, Breakdown voltage ↑, Crystallization temperature ↑, Improve scaling	Bottom Body	Interface state density ↑, Fixed charge ↑, Hysteresis ↑, BTI ↑, Mobility ↓ Trap density ↑, N pile up at the interface, Mobility ↓

gate work function materials, Ta, Hf, Zr, and Al, which are n$^+$ poly compatible, were highly reactive with gate dielectrics, SiO$_2$. TaSiN and Ru–Ta alloy were close to n$^+$ poly work function, but their work functions vary at high temperature. TaN and TiN have been good choices of high-k application because of good thermal stability, appropriate work function, and easy process. However, so far, there is no conventional CMOS compatible dual metal gate candidate without increasing a process step. Single metal gate with dual workfunction could be the ideal strategic choice of technology depending on applications [63–67].

Full silicidation of poly gate electrode could be the strong candidate for the next generation gate electrode technology [68–70]. By making full silicidation of poly Si with Co or Ni, one could possibly achieve adjustable work functions (<0.3 eV) either toward n$^+$ poly or p$^+$ poly (Fig. 1.5) and it might be the

TABLE 1.2: Role of Si and N in HfO_2 Gate Dielectrics [62]

MERITS	PROFILE	ISSUES
Impurity diffusion (B,P) ↓, Oxygen diffusion (interfacial oxide) ↓, Strong bond with N, Scaling	Top	Dielectric constant ↓, Process complexity, Reliability?
Impurity diffusion ↓, Strong bond with N, Oxygen diffusion (interfacial oxide) ↓, Crystallization temperature with N ↑ (1000°C)	Body	Dielectric constant ↓↓, Worsen scaling, N pile up at interface, Reliability?

Si traps N and suppress N out-diffusion; thus increase N%.
Si can be used to tailor the N profile.

ideal application of FDSOI (fully depleted silicon on insulator), which has low threshold voltage (Fig. 1.5). In addition, it will not increase process complexity as well.

1.2   BEYOND 45 nm TECHNOLOGY

As mentioned earlier, SOI and strained Si will be the another major change in CMOS technology. Advanced double gate structure (Fig. 1.6), FinFET, has also been studied and it has shown high drain current capability. It is also suggested that air gap BEOL (back end of line), metal gate, FinFET, high-*k*, strained Si, and FD-SOI would be the strong choice of advanced technology. In addition, new structures, new functional integration, architecture, and concept are being explored. For example, MRAM (magnetic random access memory), eDRAM (embedded dynamic RAM), and analog/mixed signal device have been developed and are developed and are being examined for replacing conventional single function devices (flash

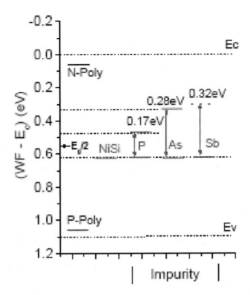

FIGURE 1.5: Work function obtained from NiSi gate electrode with dopants [69].

memory, DRAM). SoC (system on chip) and SiP (system in package) have also been introduced and these devices will be leading a semiconductor technology in near future. As technology goes beyond 45 nm node, extremely challenging issues arise such that the paradigm is changing from "everyone can do it" to "a few can do it."

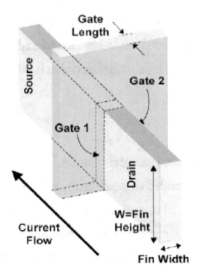

FIGURE 1.6: FinFET.

1.3   ISSUES IN HIGH-*k* DIELECTRICS

High-*k* dielectric applications include inter-poly dielectric of flash memory device to increase coupling ratio and reduce leakage current, gate dielectrics of DRAM and MIM capacitor, low-power and high-performance CMOS device. Even though lots of improvements have been made, lots of issues have also been identified. First, reliability characteristics are not fully understood and need to be focused in more detail. Second, the origin of charge trapping and degradation mechanism under dc and ac stress including TDDB (time-dependent dielectric breakdown) and BTI (bias temperature instability) are the urgent issues regarding the reliability of high-*k* dielectrics. Nature of bi-layer structure has not been thoroughly investigated in terms of reliability as well. It is also worth noting that gate leakage current of high-*k* dielectrics is still high even when it is scaled down below 8 Å. It is a potential issue, especially for low power applications. This indicates that another binary or ternary oxide with higher-*k* (25–50) must be introduced, while keep crystallization temperature high and barrier height high enough. Fermi level pining effect has been a hot issue for high-*k* dielectrics society and need to be clarified to everyone. Mobility degradation mechanism is not clear yet. Remote phonon effect and asymmetric interface density across forbidden band were suggested to explain mobility degradation for NMOSFET. As a whole, device performance may meet the criteria for real use. However, it seems that there is still lack of understanding about nature of high-*k* dielectrics. So one may have to find out what is intrinsic and what is extrinsic.

1.4   OUTLINE

Most of reliability works have been aimed to find out fundamental characteristics. In general, Chapter 2 is intended to describe the reliability characteristics of HfO_2 under ac/dc stress with regard to area dependence, thickness dependence, Weibull slope comparison, polarity dependence, and lifetime extrapolation. Soft breakdown and hard breakdown are compared in terms of Weibull distribution, acceleration factor, and lifetime projection. The breakdown behavior might be related to nature of bi-layer structure of high-k dielectrics. The chapter also discusses that by using ac stress, on/off time relation was found to have sgnificant impact on device lifetime projection. This might be attributed to not only less effective trapping during on-time, but also detrapping during off-time.

Chapter 3 discusses high temperature forming gas/D_2 annealing process related reliability characteristics. It is shown that there were trade-offs between performance and reliability. Even though one could improve performance of high-k dielectrics devices by using forming gas or D_2 annealing, it has shown little degraded reliability characteristics resulting from hydrogen-related defect generation under high field stress. Trade-offs in terms of TDDB, TZDB, BTI, Weibull slope, and lifetime projection are also discussed.

In chapter 4, we study the intensive reliability characteristics of dual metal gate on bi-layer structure of HfO_2 shown. In fact, there are several facts that low barrier height could possibly change wear-out properties of HfO_2 and its statistical value, Weibull slope. These effects are mainly due to different charge fluence and different nature of chemical and physical properties between high-k layer and interface layer.

Chapter 5 is dedicated to breakdown model of high-k system and includes bi-modal defect generation rate, charge to breakdown, critical defect density, and band modeling. The bottom line is that charge fluence is strongly dependent on barrier height, temperature, and thickness. Therefore, low barrier height of high-k layer might have impact on breakdown wear-out properties. This explains a charge fluence effect depending on barrier height and nature of bi-layer structure.

REFERENCE

[1] B. Davari, R. H. Dennard, and G. G. Shahidi, "CMOS scaling for high-performance and low-power-the next ten years," *Proc. IEEE*, Vol. 89, pp. 595–606, 1995. doi:10.1109/5.371968

[2] Y. Taur, D. Buchanan, W. Chen, D. Frank, K. Ismail, S.-H. Lo, G. Sai-Halasz, Viswanathan, H.-J. C. Wann, S. Wind, and H.-S. Wong, "CMOS

scaling into the nanometer regime," *Proc. IEEE*, Vol. 85, pp. 486–504, 1997. doi:10.1109/5.573737

[3]   S. Asai and Y. Wada, "Technology challenges for integration near and below 0.1 μm," *Proc. IEEE*, Vol. 85, pp. 505–520, 1997. doi:10.1109/5.573738

[4]   T. Sugii, Y. Momiyama, M. Deura, and K. Goto, "MOS scaling beyond 0.1 μm," in *Silicon Nanoelectronics Workshop*, June 1999, pp. 60–61.

[5]   H.-S. P. Wong, D. J. Frank, P. M. Solomon, H.-J. Wann, and J. Welser, "Nanoscale CMOS," *Proc. IEEE*, Vol. 87, pp. 537–570, 1999. doi:10.1109/5.752515

[6]   R. Yan, A. Ourmazd, and K. F. Lee, "Scaling the Si MOSFET: From bulk to SOI to bulk," *IEEE Trans. Electron Devices*, Vol. 39, pp. 1704–1710, 1992. doi:10.1109/16.141237

[7]   D. J. Frank, S. E. Laux, and M. V. Fischetti, "Monte Carlo simulation of a 30 nm dual-gate MOSFET: How far can Si go?," *IEDM Tech Dig.*, pp. 553–556, 1992.

[8]   R. W. Keyes, "The effect of randomness in the distribution of impurity atoms on FET thresholds," *Appl, Phys.*, Vol. 8, pp. 251–259, 1975. doi:10.1007/BF00896619

[9]   X. Tang, V. K. De, and J. D. Meindl, "Intrinsic MOSFET parameter fluctuation due to random dopant placement," *IEEE Tans. VLSI Syst.*, Vol. 5, pp. 369–376, 1997. doi:10.1109/92.645063

[10]   V. K. De, X. Tang, and J. D. Meindl, "Scaling limits of Si MOSFET technology imposed by random parameter fluctuations," *Proc. IEEE Device Res. Conf. Dig.*, pp. 114–115, June 1996.

[11]   Y. Yasuda, M. Takamiya, and T. Hirmoto, "Effects of impurity position distribution on threshold voltage fluctuations in scaled MOSFETs," in *Si Nanoelectronics Workshop Abstracts*, June 1999, pp. 26–27.

[12]   H.-S. Wong and Y. Taur, "Three-dimensional 'atomistic' simulation of discrete microscopic random dopant distributions effects in sub-0.1 μm MOSFETs," *IEDM Tech. Dig.*, pp. 705–708, 1993.

[13] H.-S. Wong, Y. Taur, and D. Frank, "Discrete random dopant distribution effects in nanometer-scale MOSFETs," *Microelectron. Reliability*, Vol. 38, No. 9, pp. 1447–1456, 1998. doi:10.1016/S0026-2714(98)00053-5

[14] A. Asenov and S. Saini, "Random dopant fluctuation resistant decanano MOSFET architectures," in *Si Nanoelectronics Workshop Abstracts*, June 1999, pp. 84–85.

[15] M. Saito, M. Yoshida, K. Asaka, H. Goto, N. Fukuda, M. Kawando, M. Kojima, M. Suzuki, K. Ogaya, H. Enomoto, K. Hotta, S. Sakai, H. Asakura, T. Fukuda, T. Sekiguchi, T. Takakura, and N. Kobayashi, "Advanced thermally stable silicide S/D electrodes for high-speed logic circuits with large-scale embedded Ta_2O_5-capacitor DRAMs," *IEDM Tech. Dig.*, pp. 805–808, 1999.

[16] P. K. Roy and I. C. Kizilyalli, "Stacked high-ε gate dielectric for gigascale integration of metal-oxide-semiconductor technologies," *Appl. Phys. Lett.*, Vol. 72, pp. 2835–2837, 1998. doi:10.1063/1.121473

[17] D. Park, Y.-C. King, Q. Lu, T.-J. King, C. Hu, A. Kalnitsky, S.-P. Tay, and C.-C. Cheng, "Transistor characteristics with Ta_2O_5 gate dielectric," *IEEE Electron Device Lett.*, Vol. 19. pp. 441–443, 1998. doi:10.1109/55.663533

[18] H. F. Luan, S. J. Lee, C. H. Lee, S. C. Song, Y. L. Mao, Y. L. Mao, Y. Senzaki, D. Roberts, and D. L. Kwong, "High quality Ta_2O_5 gate dielectrics with Tox,eq<10 Å," *IEDM Tech. Dig.*, pp. 141–144, 1999.

[19] A. Chatterjee, R. A. Chapman, K. Joyner, M. Otobe, S. Hattangady, M. Bevan, G. A. Brown, H. Yang, Q. He, D. Rodgers, S. J. Fang, R. Kraft, A. L. P. Rotondaro, M. Terry, K. Brennan, S.-W. Aur, J. C. Hu, H.-L. Tsai, P. Jones, G. Wilk, M. Aoki, M. Rodder, and I.-C. Chen, "CMOS metal replacement gate transistors using tantalum pentoxide gate insulator," *IEDM Tech. Dig.*, pp. 777–780, 1998.

[20] H.-S. Kim, D. C. Gilmer, S. A. Campbell, and D. L. Polla, "Leakage current and electrical breakdown in metal-organic chemical vapor deposited TiO_2 dielectrics on silicon substrates," *Appl. Phys. Lett.*, Vol. 69, pp. 3860–3862, 1996. doi:10.1063/1.117129

[21] B. H. Lee, Y. Jeon, K. Zawadzki, W.-J. Qi, and J. C. Lee, "Effects of interfacial layer growth on the electrical characteristics of thin titanium oxide films on silicon," *Appl. Phys. Lett.*, Vol. 74, pp. 3143–3145, 1999. doi:10.1063/1.124089

[22] C. Hobbs, R. Hedge, B. Maiti, H. Tseng, D. Gilmer, P. Tobin, O. Adetutu, F. Huang, D. Weddington, R. Nagabushnam, D. O'Meara, K. Reid, L. La, L. Grove, and M. Rossow, "Sub-quarter micron CMOS process for TiN-gate MOSFETs with TiO_2 gate dielectric formed by titanium oxidation," *Symp. VLSI Tech. Dig.*, 1999, pp. 133–134.

[23] K. J. Hubbard and D. G. Schlom, "Thermodynamic stability of binary oxides in contact with silicon," *J. Mat. Res.*, Vol. 11, pp. 2757–2776, 1996.

[24] R. D. Shannon, "Dielectric polarizabilities of ions in oxides and fluorides," *J. Appl. Phys.*, Vol. 73, pp. 348–366, 1993. doi:10.1063/1.353856

[25] J. Robertson, "Band offsets of wide-band-gap oxides and implications for future electronic devices," *J. Vac. Sci. Tech. B*, Vol. 18, pp. 1785–1791, 2000. doi:10.1116/1.591472

[26] W.-J. Qi, R. Nieh, B. H. Lee, L. Kang, Y. Jeon, K. Onishi, T. Ngai, S. Banerjee, and J. C. Lee, "MOSCAP and MOSFET characteristics using ZrO_2 gate dielectric deposited directly on Si," *IEDM Tech. Dig.*, pp. 145–148, 1999.

[27] B. H. Lee, L. Kang, W.-J. Qi, R. Nieh, Y. Jeon, K. Onishi, and J. C. Lee, "Ultrathin hafnium oxide with low leakage and excellent reliability for alternative gate dielectric application," *IEDM Tech. Dig.*, pp. 133–136, 1999.

[28] W.-J. Qi, R. Nieh, B. H. Lee, K. Onishi, L. Kang, Y. Jeon, J. C. Lee, V. Kaushik, B.-Y. Neuyen, L. Prabhu, K. Eigenbeiser, and J. Finder, "Performance of MOSFETs with ultra thin ZrO_2 and Zr silicate gate dielecrics," *Symp. VLSI Tech. Dig.*, pp. 40–41, 2000.

[29] B. H. Lee, R. Choi, L. Kang, S. Gopalan, R. Nieh, K. Onishi, and J. C. Lee, "Characteristics of TaN gate MOSFET with ultra thin hafnium oxide," *IEDM Tech. Dig.*, pp. 39–42, 2000.

[30] L. Kang, Y. Jeon, K. Onishi, B. H. Lee, W.-J. Qi, R. Nieh, S. Gopalan, and J. C. Lee, "Single-layer thin HfO$_2$ gate dielectric with n+-polysilicon gate," *Symp. VLSI Tech. Dig.*, pp. 44–45, 2000.

[31] C. H. Lee, H. F. Luan, W. P. Bai, S. J. Lee, T. S. Jeon, Y. Senzaki, D. Roberts, and D. L. Kwong, "MOS characteristics of ultra thin rapid thermal CVD ZrO$_2$ and Zr silicate gate dielectrics," *IEDM Tech. Dig.*, pp. 27–30, 2000.

[32] M. Koyama, K. Suguro, M. Yoshiki, Y. Kamimuta, M. Koike, M. Ohse, C. Hongo, and A. Nishiyama, "Thermally stable ultra-thin nitrogen incorporated ZrO$_2$ gate dielectric prepared by low temperature oxidation of ZrN," in *IEDM Tech. Dig.*, pp. 459–462, 2001.

[33] R. Nieh, S. Krishnan, H.-J. Cho, C. S. Kang, S. Gopalan, K. Onishi, R. Choi, and J. C. Lee, "Comparison between ultra-thin ZrO$_2$ and ZrOxNy gate dielectrics in TaN or poly-gated NMOSCAP and NMOSFET devices," *Symp. VLSI Tech. Dig.*, pp. 186–187, 2002.

[34] M. T. Thomas, "Preparation and properties of sputtered hafnium and anodic HfO$_2$ films," *J. Electrochem. Soc.*, Vol. 117, pp. 396–403, 1970.

[35] K. Kukli, J. Ihanus, M. Ritala, and M. Leskela, "Tailoring the dielectric properties of HfO$_2$-Ta$_2$O$_5$ nanolaminates," *Appl. Phys. Lett.*, Vol. 68, pp. 3737–3739, 1996. doi:10.1063/1.115990

[36] D. Hisamoto, T. Kaga, Y. Kawamoto, and E. Takeda, "A fully depleted lean-channel transistor (DELTA)—A novel vertical ultra thin SOI MOSFET," *IEDM Tech. Digest*, p. 833, 1989.

[37] D. Fried, A. Johnson, E. Nowak, J. Rankin, and C. Willets, "A sub-40 nm body thickness N-Type FinFET," in *Proc. Device Res. Conf.*, 2001, p. 24.

[38] N. Lindert, L. Chang, Y.-K. Choi, E. Anderson, W.-C. Lee, T.-J. King, J. Bokor, and C. Hu, "Sub-60-nm quasi- planar FinFETs fabricated using a simplified process," *IEEE Electron Device Lett.* Vol. 22, pp. 487–489, 2001. doi:10.1109/55.954920

[39]   J. Kedzierski, D. M. Fried, E. J. Nowak, T. Kanarsky, J. H. Rankin, H. Hanafi, W. Natzle, D. Boyd, Y. Zhang, R. A. Roy, J. Newbury, C. Yu, Q. Yang, P. Saunders, C. P. Willets, A. Johnson, S. P. Cole, H. E. Young, N. Carpenter, D. Rakowski, B. A. Rainey, P. E. Cottrell, M. Ieong, and H.-S. P. Wong, "High-performance symmetric-gate and CMOS-compatible *V*t asymmetric-gate FinFET devices," *IEDM Tech. Digest*, pp. 437–440, 2001.

[40]   Y.-K. Choi, N. Lindert, P. Xuan, S. Tang, D. Ha, E. Anderson, T.-J. King, J. Bokor, and C. Hu, "FinFET process technology for nanoscale CMOS," *IEDM Tech Digest*, p. 421, 2001.

[41]   Vogelsang and H. R. Hofmann, "Electron transport in strained silicon layers on Si1_*x*Ge*x* substrates," *Appl Phys. Lett.* Vol. 63, p. 186, 1993. doi:10.1063/1.110394

[42]   D. Nayak, J. Woo, J. Park, K. Wang, and K. MacWilliams, "High-mobility p-channel metal-oxide-semiconductor field-effect transistors on strained Si," *Appl. Phys. Lett.*, Vol. 62, pp. 2853–2855, 1993. doi:10.1063/1.109205

[43]   J. Welser, J. Hoyt, S. Takagi, and J. Gibbons, "Strain dependence of the performance enhancement in strained-Si n-MOSFETs," *IEDM Tech. Dig.t*, pp. 373–376, 1994.

[44]   M. Fischetti and S. Laux, "Band structure, deformation potentials, and carrier mobility in strained Si, Ge, and SiGe alloys," *J. Appl. Phys.*, Vol. 80, p. 2234, 1996. doi:10.1063/1.363052

[45]   S. Tiwari, M. Fischetti, P. Mooney, and J. Welser, "Hole mobility improvement in silicon-on-insulator and bulk silicon transistors using local strain," *IEDM Tech Dig.*, pp. 939–941, 1997. doi:full_text

[46]   K. Rim, J. Hoyt, and J. Gibbons, "Transconductance enhancement in deep submicron strained-Si n-MOSFETs," *IEDM Tech. Dig.*, p. 707, 1998.

[47]   T. Mizuno, N. Sugiyama, H. Satake, and S. Takagi, "Advanced SOI-MOSFETs with strained-Si channel for high speed CMOS—Electron/hole mobility enhancement," *Symp. VLSI Technol Dig. Technical Papers*, p. 210, 2000.

[48] K. Ismail, "Si/SiGe CMOS: Can it extend the lifetime of Si," *ISSCC Tech. Dig.*, pp. 116–117, 1997.

[49] P. Mooney, "Strain relaxation and dislocations in SiGe/Si structures," *Mater. Sci. Eng.*, Vol. R17, pp. 105–146, 1996.

[50] E. Fitzgerald, Y. Xie, D. Monroe, P. Silverman, J. Kuo, A. Kortan, F. Theil, and B. Weir, "Relaxed GexSi1_x structures for III/V integration with Si and high mobility two-dimensional electron gases in Si," *J. Vac. Sci. Technol. B*, Vol. 10, p. 1087, 1992. doi:10.1116/1.586204

[51] K. Ismail, F. Nelson, J. Chu, and B. Meyerson, "Electron transport properties of Si/SiGe heterostructures: Measurements and device implications," *Appl. Phys. Lett.*, Vol. 63, pp. 660–662, 1993. doi:10.1063/1.109949

[52] K. Rim, S. Koester, M. Hargrove, J. Chu, P. M. Mooney, J. Ott, T. Kanarsky, P. Ronsheim, M. Ieong, A. Grill, and H.-S. P. Wong, "Strained Si NMOSFETs for high performance CMOS technology," *Symp. VLSI Technol Dig. Technical Papers*, p. 59, 2001.

[53] L.-J. Huang, J. Chu, C. Canaperi, C. D'Emic, R. Anderson, S. Koester, and H.-S. P. Wong, "SiGe-on-insulator prepared by wafer bonding and layer transfer for high-performance field-effect transistors," *Appl. Phys. Lett.*, Vol. 78, p. 1267, 2001. doi:10.1063/1.1371967

[54] L.-J. Huang, J. Chu, S. A. Goma, C. D'Emic, S. J. Koester, D. F. Canaperi, P. M. Mooney, S. A. Cordes, J. L. Speidell, R. M. Anderson, and H.-S. P. Wong, "Carrier mobility enhancement in strained Si-on-insulator fabricated by wafer bonding," *Symp. VLSI Technol.Dig. Technical Papers*, p. 57, 2001.

[55] Z.-Y. Cheng, M. Currie, C. Leitz, G. Taraschi, E. Fitzgerald, J. Hoyt, and D. Antoniadis, "Electron mobility enhancement in strained-Si n-MOSFETs fabricated on SiGe-on-insulator (SGOI) substrates," *IEEE Electron Device Lett.*, Vol. 22, pp. 321–323, 2001. doi:10.1109/55.930678

[56] T. Mizuno, N. Sugiyama, A. Kurobe, and S. Takagi, "Advanced SOI p-MOSFETs with strained-Si channel on SiGe-on-insulator substrate fabricated by SIMOX technology," *IEEE Trans. Electron Devices*, Vol. 48, pp. 1612–1618, 2001. doi:10.1109/16.936571

[57] T. Tezuka, N. Sugiyama, and S. Takagi, "Fabrication of strained Si on an ultrathin SiGe-on-insulator substrate with a high-Ge fraction," *Appl. Phys. Lett.*, Vol. 79, pp. 1798–1800, 2001. doi:10.1063/1.1404409

[58] T. Mizuno, N. Sugiyama, T. Tezuka, and S. Takagi, "Novel fabrication technique for relaxed SiGe-on-insulator substrates without thick SiGe buffer structures," in *Proc. Int. Conf. Solid State Devices and Materials (SSDM)*, 2001, pp. 242–243.

[59] A. R. Powell, S. S. Iyer, and F. K. LeGoues, "New approach to the growth of low dislocation relaxed SiGe material," *Appl. Phys. Lett.*, Vol. 64, pp. 1856–1858, 1994. doi:10.1063/1.111778

[60] H.-S. P. Wong, "Beyond the conventional transistor," *IBM J. Res. Dev.*, Vol. 46, No. 2/3, pp. 133–168, 2002.

[61] B. Doris, M. Ieong, T. Kanarsky, Y. Zhang, R. A. Roy, O. Dokumaci, Z. Ren, F.-F. Jamin, L. Shi, W. Natzle, H.-J. Huang, J. Mezzapelle, A. Mocuta, S. Womack, M. Gribelyuk, E. C. Jones, R. J. Miller, H.-S. P. Wong, and W. Haensch, "Extreme scaling with ultra-thin Si channel MOSFETs," in *IEDM Tech. Dig.*, pp. 267–270, 2002.

[62] SRC FEP Meeting at the NC State University in Nov. 2003.

[63] Y. Abe, T. Oishi, K. Shiozawa, Y. Tokuda, and S. Satoh, "Simulation study on comparison between metal gate and ploysilicon gate for sub-quarter-micron MOSFETs," *IEEE Electron Device Lett.*, Vol. 20, pp. 632–634, 1999. doi:10.1109/55.806111

[64] Q. Lu, R. Lin, P. Ranade, T.-J. King, and C. Hu, "Metal gate workfunction adjustment for future CMOS technology," *Symp. VLSI Tech. Dig.*, pp. 45–46, 2001.

[65] Y.-S. Suh, G. Heuss, H. Zhong, S.-N. Hong, and V. Misra, "Electrical characteristics of TaSiN gate electrode for dual gate Si-CMOS devices," *Symp. VLSI Tech. Dig.*, pp. 47–48, 2001.

[66] H. Zhong, S.-N. Hong, Y.-S. Suh, H. Lazar, G. Heuss, and V. Misra, "Properties of Ru-Ta alloy as gate electrodes for NMOS and PMOS silicon devices," *IEDM Tech. Dig.*, pp. 467–470, 2001.

[67] D.-G. Park, T.-H. Cha, K.-Y. Lim, H.-J. Cho, T.-K. Kim, S.-A. Jang, Y.-S. Suh, V. Misra, I.-S. Yeo, J.-S. Roh, J. W. Park, and H.-K. Yoon, "Robust ternary metal gate electrodes for dual gate CMOS devices," *IEDM Tech. Dig.*, pp. 671–674, 2001.

[68] W. P. Maszara, Z. Krivokapic, P. King, J.-S. Goo, and M.-R. Lin, "Transistors with dual work function metal gates by single full silicidation (FUSI) of polysilicon," *IEDM Tech. Dig.*, pp. 367–370, 2002.

[69] J. Kedzierski, D. Boyd*, P. Ronsheim*, S. Zafar, J. Newbury, J. Ott, C. Cabral Jr., M. Ieong*, and W. Haensch, "Threshold voltage control in NiSi-gated MOSFETs through silicidation induced impurity segregation (SIIS)," *IEDM Tech. Dig.*, pp. 315–318, 2003.

[70] H. Huang, D. S. Yu, A. Chin, C. H. Wu, W. J. Chen, C. Zhu, M. F. Li, B. Jin, and D.-L. Kwong , "Fully silicided NiSi and germanided NiGe dual gates on SiO_2/Si and Al_2O_3/Ge-on-insulator MOSFETs," *IEDM Tech. Dig.*, pp. 319–322, 2003.

CHAPTER 2

Hard- and Soft-Breakdown Characteristics of Ultrathin HfO$_2$ Under Dynamic and Constant Voltage Stress

2.1 MOTIVATION FOR HIGH-k GATE DIELECTRICS

In the past few decades, improvements in speed and shrinkage of chip area of integrated circuits have been achieved by scaling down of physical thickness of the SiO$_2$ gate dielectrics and design rule. Driving technology of modern integrated circuits greatly relies on aggressive scaling down of MOS transistors. Reducing the transistor size not only increases package densities, but also increases circuit speed and reduces power dissipation [1]. However, as technology evolves, SiO$_2$ will soon reach its physical limitation such as high leakage current and reliability concerns. Continuing scaling down of the MOSFET device with the minimum feature size of 90 nm and below would require EOT (equivalent oxide thickness) of less than 15 Å. A 10–15-Å thick SiO$_2$ layer corresponding to only around 3–4 monolayers of SiO$_2$. In this thinner EOT range, SiO$_2$ suffers from leakage current and it is too high to be used, particularly for low-power operations. In addition, SiO_2 thickness control across a 12-inch wafer imposes even more serious difficulty in the growth of such a thin film. Even a monolayer difference in thickness represents a large percentage difference and thus can result in varying device performance parameters across the wafer [2]. Moreover, reliability also becomes a more challenging issue for a SiO_2 film of only 10–15-Å thickness [3]. In such a thin dielectrics regime, soft breakdown behaviors as well as hard breakdown need to be considered. In order to satisfy the strict requirement, SiO_2 thickness must be reduced below 10 Å in the next 5 years. In this thickness range, direct tunneling current through SiO_2 cannot be tolerated anymore. Therefore, it has become necessary to identify alternative high-*k* gate dielectrics in order to meet the stringent requirements for leakage current and EOT. So far, a number of high-*k* dielectrics have been reported such as HfO_2, ZrO_2, Al_2O_3, and their silicates [4–9]. Among various high-*k* gate dielectric materials, HfO_2 and its silicates with moderate high-*k* value ($\varepsilon = 21-25$) have been found to be attractive materials because they have demonstrated good device characteristics and are compatible with the polysilicon gate process as well as metal gates [10, 11]. Fig. 2.1 shows the band alignments of various high-*k* materials with respect to silicon [12]. As can be seen in this figure, HfO_2 shows adequate band gap (>5 eV) and band offsets (>1 eV).

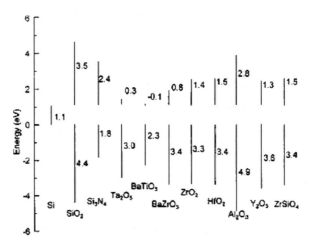

FIGURE 2.1: Band alignments of high-*k* materials with respect to silicon.

2.2   RELIABILITY ISSUES OF HIGH-k DIELECTRICS

High-k dielectrics have been known to be "trap-rich" materials. A number of efforts were carried out to improve the characteristics of high-k dielectrics. For example, high temperature deuterium and forming gas anneals and O$_3$ surface treatments have been shown to reduce interface state [13–15]. The origin of traps in high-k dielectrics, however, still remains a question. These preexisting traps may play an important role in dielectric wear-out as well as device performance. In order to evaluate reliability, it is necessary to pinpoint the factors that influence the breakdowns of HfO$_2$. The parameters that are experimentally measured to evaluate the reliability of MOS devices must be carefully collected by statistical methods. This chapter addresses issues such as soft breakdown, hard breakdown, the physical and chemical nature of the interface layer and the bulk high-k layer, bimodal defect generation rates, polarity dependence due to asymmetric band structure, charging effect by preexisting traps, thickness dependence, area scaling, and ac stressing.

2.3   BREAKDOWN BEHAVIORS OF HfO$_2$ UNDER dc STRESSING

2.3.1   HfO$_2$ Fabrication and Experiments

HfO$_2$ was deposited using reactive dc magnetron sputtering with O$_2$ modulation technique [10], followed by postdeposition annealing (PDA) at 500°C. Polysilicon (200-nm thick) was deposited at 580°C in an amorphous state using LPCVD, followed by photolithography/etch process. Phosphorous was implanted at 50 keV with a dose of 5×10^{15} cm^{-2} for NMOS devices. Source and drain (S/D) oxidation at 850°C for 1 min was performed, followed by low-temperature oxide deposition and contact patterning. Dopant activation was carried out at 950°C for 1 min. Sputtered aluminum was used for both interconnect and backside metallization. Various sizes of capacitors were used for TDDB evaluation. The devices having initial gross defects (early failures) were screened out by $C–V$ and leakage current measurements at low voltage using a HP4194A LCR meter and a HP 4156 semiconductor parameter analyzer, respectively.

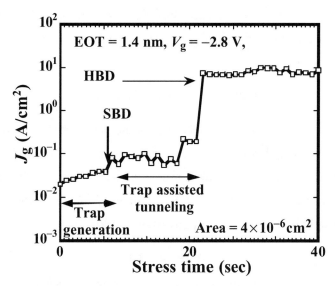

FIGURE 2.2: Typical breakdown behavior of HfO_2 MOSCAP.

2.3.2 Soft Breakdown of HfO_2

Fig. 2.2 shows the typical breakdown behavior of HfO_2 (EOT = 1.4 nm, physical thickness = 4.8–5 nm) gate dielectrics, which consists of soft breakdown and hard breakdown.

In general, soft breakdown is considered to result from a weak localized percolation path between the gate electrode and the substrate. A critical number of electron traps are generated in the gate dielectric layer and at the interface, which in turn form percolative clusters [16–19]. Fluctuation of the leakage current in the soft breakdown region results from the trapping–detrapping of electrons in the percolation clusters. The soft breakdown behavior is believed to be similar to that of ultrathin SiO_2 gate dielectrics. However, the amount of leakage current after soft breakdown is quite different. In the case of SiO_2 the current increase after the onset of soft breakdown depends on the device size. The radius of soft breakdown path of high-k dielectrics or the origin of soft breakdown can be very different since high-k dielectrics are in general bilayer structures (an interface and a bulk high-k layer). Soft breakdown of HfO_2 (EOT = 1.4 nm) has been predominantly observed as the first breakdown event in the experiment (stress voltage from −2.6 to −2.8 V), and the time between soft and hard breakdown significantly decreases as stress voltage increases. The statistics of gate oxide breakdown are usually described by the Weibull distribution:

$$F(x) = 1 - \exp[-(x/\alpha)\beta],$$

where β is the Weibull slope, which is an important parameter when evaluating gate oxide reliability. Weibull slope, β, is also useful in predicting lifetime distribution for different capacitor areas. For example, if the area is increased by a factor of (A/A'), then the distribution is shifted by a factor of $\ln(A/A')$, and the characteristic lifetime α would be decreased to α', according to [20]

$$(\alpha/\alpha') = (A/A')^{1/\beta}.$$

Fig. 2.3 shows the Weibull distributions of HfO$_2$ (EOT = 1.4 nm) for both soft and hard breakdown. Weibull slopes of soft and hard breakdown are of different

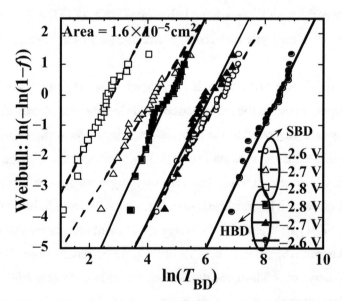

FIGURE 2.3: Weibull distribution comparison between soft and hard breakdown of HfO$_2$. Solid lines represent the hard breakdown distributions, and the dashed lines represent the soft breakdown. The Weibull slope, β, of hard breakdown is about 2, whereas that of soft breakdown is about 1.4. (EOT = 1.4 nm).

values. For hard breakdown, Weibull slope was found to be about 2 and that of soft breakdown was about 1.4. The pattern is similar to that of soft and progressive breakdown of SiO_2, which also exhibit different β values [21]. In the case of SiO_2, the first breakdown could be either soft or hard, but both breakdown events have β values that are similar to those of HfO_2 and thus a common origin in statistical characteristics [22]. Furthermore, for SiO_2, it has been reported that Weibull distribution of the first breakdown always shows lower values compared to the second one [23]. However, for HfO_2 (EOT $= 1.4$ nm), the soft breakdown predominantly occurs as the first breakdown event and β values for soft breakdown and hard breakdown are quite different. The difference could simply be due to statistical effects.

However, a study on another high-k structure (Ta_2O_5/SiO_2 stack) suggested that the high voltage breakdown of the dielectric stack was completely determined by the interfacial SiO_2 layer. This is due to the high electric field across the interfacial layer, which in turn leads to bulk Ta_2O_5 breakdown immediately after interface degradation [24]. Similarly, the difference in the β values of soft and hard breakdowns of HfO_2 may also be due to the interface effects. That is, the charge trapping within the interfacial layer not only triggers soft breakdown of HfO_2 but also influences its Weibull distribution. The charge fluences and electric field across the interfacial layer are much larger than that across the bulk HfO_2 layer under substrate injection, and there are different defects generation mode as well as different charge fluences between bulk HfO_2 and interface layer under gate injection. Obviously, these differences depend on the composition and the thickness of the interface layer. In this experiment, we did not intentionally deposit an interface layer prior to high-k deposition, but it grew as a result of subsequent thermal processing (below 10 Å). If the interface layer is made thicker (more than 10 Å), part of the voltage that drops across the interface layer during the stressing will drop across the HfO_2 after interface degradation. This possibly results in hard breakdown under substrate injection in thicker dielectrics. Based on the thickness dependence and the percolation model, it is thought that defect generation and charge trapping in bulk HfO_2 affect hard breakdown. On the other hand, EOT typically increases

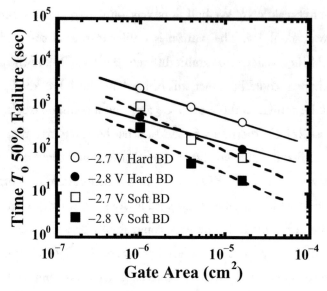

FIGURE 2.4: Gate area scaling of HfO$_2$ in terms of soft and hard breakdown. Different gate area scaling behavior are shown between soft and hard breakdown (EOT = 1.4 nm).

due to the growth of the interfacial layer during the subsequent processing, and β has a strong thickness and a defect density dependence [16]. Therefore, it is necessary to investigate the interfacial layer in detail as a part of gate dielectrics. In addition, the results indicate that separate detection and analysis of soft and hard breakdown are necessary in order to give a more precise breakdown distribution as well as lifetime projection of HfO$_2$. Area dependence of time-dependent dielectric breakdown of HfO$_2$ (EOT = 1.4 nm) is shown in Fig. 2.4. It can be seen that the area dependences for both soft and hard breakdown correlate with the Weibull distribution with the respective β values. In other words, lifetime of HfO$_2$ can be projected to different capacitor sizes using the appropriate β value. The different physical and chemical nature of the interfacial layer and the bulk layer of HfO$_2$ [25] may induce different breakdown behaviors. The defect generation rate and the critical defect density have a strong dependence on voltage and thickness, respectively [18, 19]. The interfacial layer of HfO$_2$ is physically thinner than the bulk HfO$_2$ and is exposed to high electric field due to its low dielectric constant.

Furthermore, HfO_2 gate stack will go through various tunneling conductions and defect generations under the different polarities. Thus, the role that the interfacial layer play in triggering soft breakdown can be higher defect generation rate due to high stress field or lower critical defect density due to thinner thickness. We assume that two homogeneous isotropic dielectric media with different dielectric constants, high dielectric constant ε_1 and low dielectric constant ε_2, are separated by a charge-free boundary. We have

$$E_2 = E_1(\varepsilon_1/\varepsilon_2),$$

where E is an electric field [26]. The equation indicates that the internal electric field is dependent on the dielectric constant. A wide range of t_{BD} ($>10^4$ s) has been analyzed for long-term reliability study and to investigate the precise lifetime projection and voltage acceleration factors (Fig. 2.5). Gate voltage to assure a 10-year lifetime is also shown in Fig. 2.5. The data shown in lifetime projection also has $\beta = 2$, which is same as that shown in Fig. 2.3. Estimated 10-year lifetime

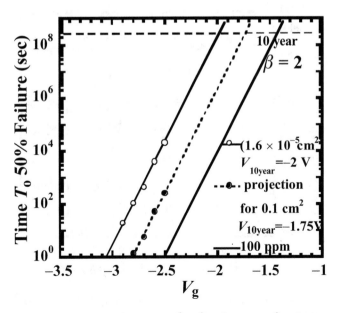

FIGURE 2.5: A 10-year lifetime of 1.6×10^{-5} cm^2 and its 0.1 cm^2 and 100 ppm projection of HfO_2 with $\beta = 2$.

operating voltage has been projected to be -2 V. Projection for 0.1 cm^2 and 100 ppm have been estimated to be -1.75 V and -1.4 V using the obtained Weibull value of $\beta = 2$.

2.3.3   Thickness Dependence of HfO₂ Hard Breakdown

So far, we have explained the area dependence of the breakdown distribution, Weibull slopes, and scaling factors for both soft and hard breakdown [27]. Kauerauf *et al.* reported that based on the analysis of Weibull slope for Al₂O₃ as a function of physical thickness, breakdown in high-k layer is dominated by an extrinsic mechanism [28]. Kerber *et al.* reported that the thickness dependence of wear-out properties of Al₂O₃ tend to support intrinsic breakdown mechanism [29]. ZrO₂, however, shows weak thickness dependences [30]. Thus, the thickness dependence of hard breakdown of high-k dielectrics is not well understood.

Fig. 2.6 shows the breakdown voltage distributions of different EOT thickness (1.4 and 2.5 nm).

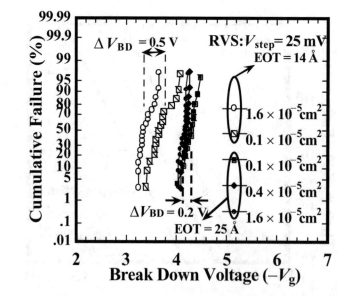

FIGURE 2.6: Breakdown voltage distributions of HfO₂ MOSCAP (EOT $= 1.4$ and 2.5 nm) by ramped voltage stress measurement (RVS). Wider distribution can be observed for devices with EOT $= 1.4$ nm.

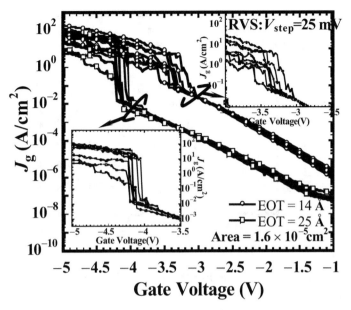

FIGURE 2.7: Narrower distribution and more abrupt breakdown behavior was observed for thicker HfO$_2$ (EOT = 2.5 nm) by ramp voltage stress. The insets emphasize the breakdown characteristics.

HfO$_2$ with EOT of 2.5 nm has higher breakdown voltage, but lower breakdown field than HfO$_2$ with EOT = 1.4 nm ($T_{\text{interface}}$ < 0.9 nm). And similar to SiO$_2$, the breakdown voltage distribution of the thinner HfO$_2$ device (1.4 nm) is wider than that of thicker films (2.5 nm). Also, like SiO$_2$, thicker HfO$_2$ shows more abrupt breakdown characteristics compared to thinner HfO$_2$ (Fig. 2.7).

For SiO$_2$, a strong thickness dependence of Weibull slope has been observed under single defect generation mode and can be explained by the percolation model [16]. In this section, we show the reliability characteristics of various thickness of HfO$_2$ gate stack (EOT = 1.4–2.5 nm) using constant voltage stress (CVS) along with Weibull slope behavior. Fig. 2.8 plots the area-scaled Weibull distribution, where TDDB results of four different capacitor sizes (A_x) are normalized to those of 4.0×10^{-6} cm^2 devices (A_{norm}) by multiplying the values with $\ln(A_x/A_{\text{norm}})$.

The fact that the normalized breakdown distributions of various capacitor areas merge to a single line suggests that the breakdown is intrinsic and can be

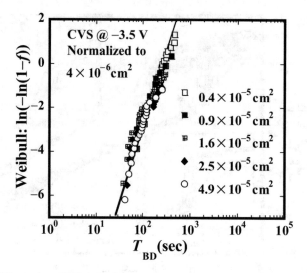

FIGURE 2.8: Normalized Weibull distributions of 2.5 nm HfO₂ with various capacitor areas by CVS.

explained by the percolation model. The slope of the Weibull distribution is an important factor in reliability prediction, where it is used for extrapolating lifetime to different percentiles and capacitor sizes. Fig. 2.9 plots the time-to-breakdown (50%) versus capacitor area as a function of HfO₂ thickness. It can be seen that thinner oxide exhibits a larger area scaling factor $(1/\beta)$.

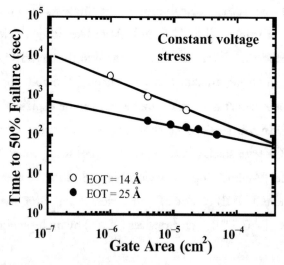

FIGURE 2.9: Time-to-breakdown (50%) versus capacitor area as a function of HfO₂ thickness. Thicker oxide exhibits smaller area scaling factor $(1/\beta)$.

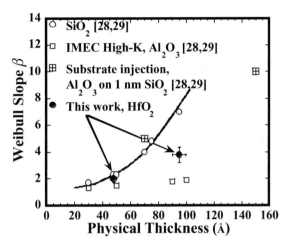

FIGURE 2.10: Weibull slope comparison between SiO_2, Al_2O_3, and HfO_2.

The trend is similar to that for SiO_2, i.e. Weibull slope decreases as thickness decreases. However, the magnitudes of β are quite different from those of SiO_2. For SiO_2, $\beta \sim 8$ is expected for intrinsic breakdown [19], while HfO_2 of similar physical thickness ($T_{phy} = 9.5$–9.8 nm, EOT = 2.5 nm) has a smaller β (< 4). The percolation model for dielectric breakdown can also be used to explain the thickness dependence of Weibull distribution [16]. For gate injection, our data show that the Weibull slope, $\beta \approx 2.0$, 4.0 for EOT of 1.4 nm and 2.5 nm, respectively (Fig. 2.10).

It was reported that the charge trapping and the defect generation exhibit a strong polarity dependence of the electrical stress in high-k gate stack (e.g. Al_2O_3) [29]. For substrate injection, it was proposed that the reliability is limited by electron trap generation in the bulk of Al_2O_3 rather than in the thin SiO_2 interfacial layer, thus strong thickness dependence of the β values was observed, as expected from the percolation model [29]. For gate injection, on the other hand, it was also suggested that the breakdown of Al_2O_3 is determined by process-induced defects causing weak spot in oxide [28]. But it has also been suggested that high voltage breakdown of thick Ta_2O_5/SiO_2 stack (60 Å) is completely determined by the interfacial SiO_2 layer because of high electric field at the interface. This causes Ta_2O_5 to breakdown immediately after interface degradation [24]. In fact, there is no universal model that can explain all previous results. In other words, high-k system may require considerations different from SiO_2, e.g., bimodal defect generations due to different physical nature of both interface and bulk layer, voltage drop as a function of interface layers, different charge fluences by different polarities, and critical defect density for breakdown, which varies with thickness. In general, high-k dielectrics might consist of an interfacial layer, various microstructures within the film, and Al or N incorporation. Therefore, one may expect that the reliability and breakdown characteristics are far more complicated than SiO_2 because of the aforementioned fact. The lower β values observed for HfO_2 (in comparison to SiO_2) might be explained by aforementioned factors including the breakdown dependence on defect density (N_{BD}) [31] and defect size (a_0) [32]. In comparison to SiO_2, HfO_2 has smaller critical defect density (N_{BD}) at breakdown and/or larger spacing between defects where tunneling of a trapped electron becomes probable. This results in a smaller Weibull slope compared to SiO_2 [33].

2.4   DYNAMIC RELIABILITY OF HfO₂

Most gate dielectric reliability testing makes use of dc stressing because of simplicity. However, in an actual circuit, gate dielectrics are, in general, under dynamic stressing. It has been reported that dynamic stressing of SiO_2 led to higher time-to-breakdown than static dc stressing because of the reduction of charge trapping in the dielectrics [34–37]. However, most reliability studies of HfO_2 have been done with static dc stressing. In this work, a detailed study on ac stressing has been performed.

2.4.1 Definition and Characteristics

Fig. 2.11 shows the waveform of unipolar stress and its definition.

The term on-time to 50% failure means that the breakdown time was counted on the basis of only the on-time to the failure (t_{on}). For SiO_2, ac stressing with respect to frequency was investigated at room temperature as well as at high temperature

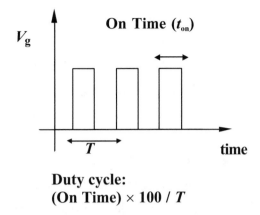

FIGURE 2.11: Wave form of unipolar stress and its definition.

conditions [36, 37]. In this experiment, we have used unipolar stressing and then did comparison with dc stressing.

The leakage current characteristics of HfO$_2$ used in this study are well controlled and uniform (Fig. 2.12). Basically, HfO$_2$ consists of bulk high-k layer and interface layer. The interface layer thickness due to subsequent thermal processing was well controlled and very thin (ranging from 5 to 9 Å; based on TEM image, data

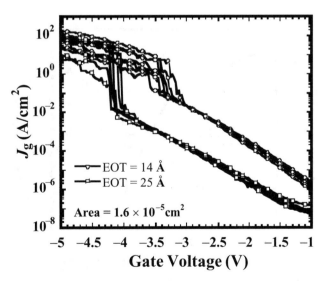

FIGURE 2.12: Leakage current comparisons between HfO$_2$ with EOT = 1.4 and 2.5 nm used in this study.

not shown). HfO$_2$ with EOT of 2.5 nm (T_{ph} = 97 Å) has higher breakdown volt-age, but lower breakdown field than that of EOT = 1.4 nm (T_{ph} = 48 Å). Similar to SiO$_2$, the breakdown voltage distribution of the thinner HfO$_2$ device (1.4 nm) is wider than that of thicker films (2.5 nm). Also, like SiO$_2$, thicker HfO$_2$ shows more abrupt breakdown characteristics compared to thinner HfO$_2$ (Fig. 2.12).

2.4.2 Frequency and Duty Cycle Dependence

Fig. 2.13 indicates that unipolar stressing on HfO$_2$ results in a longer lifetime of MOS devices with respect to duty cycles and frequencies. It has been shown previously that dynamic stressing of SiO$_2$ increased the time-to-breakdown [34, 35] and reduced interface trap density (D_{it}) generation [36, 37]. It has been suggested that the enhancement in lifetime observed under dynamic stress conditions occurs only for electric fields and oxide thickness where charge trapping is significant [37].

In fact, charge trapping rate cannot keep up with voltage transition in unipo-lar wave. The charge trapping characteristics depend on the dielectric thickness. Therefore, it is believed that as the thickness decreases (corresponding to direct

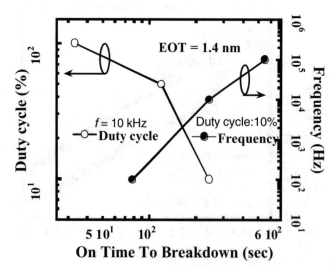

FIGURE 2.13: Smaller duty cycle and higher frequency stressing increase time to break-down.

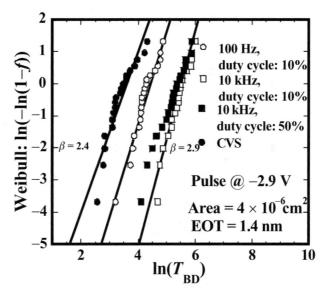

FIGURE 2.14: Longer breakdown time T_{BD} have been observed under different frequency unipolar stress.

tunneling current increase), the enhancement of the time to breakdown from ac stress is expected to be reduced. The Weibull slope of HfO_2 under unipolar stressing is slightly higher than that under dc stress (Fig. 2.14).

Low β of HfO_2 with the same physical thickness compared to that of SiO_2 would result in reduction of the lifetime scaling to percentile and total oxide area on chip. However, for the practical application, ultrathin EOT (<2 nm) of HfO_2 will be used. Thus, β value (>2) of HfO_2 is actually larger than that of SiO_2 with the similar EOT value (SiO_2: $\beta < 1.3$–1.4 for EOT = 1.4 nm) [16, 32, 38] and therefore lifetime scaling might be better than that of SiO_2 with the similar EOT value (Fig. 2.15). Note that reducing duty cycles for the ac stress also enhances the lifetime (Fig. 2.14).

2.4.3 Thickness Dependence of Dynamic Reliability

In Fig. 2.16, a larger lifetime enhancement ($t_{50(pulse)}/t_{50(dc)}$) is observed as thickness increases under the same frequency and duty cycle. Similar observation was made for SiO_2 [37]. This thickness dependence of the lifetime enhancement might result

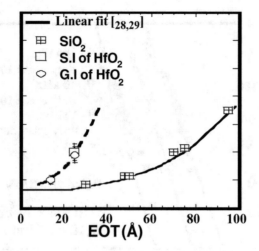

FIGURE 2.15: Weibull slope comparison with respect to EOT.

from different amount charge trapping. As thickness decreases, direct tunneling becomes dominant, which leads to relatively less charge trapping for thin dielectrics under ac stressing. Note that in Fig. 2.16, the stress voltages are different for the two samples (14 and 25 Å).

In addition, thicker HfO$_2$ shows enhanced frequency dependence of time-to-breakdown (Fig. 2.17). It has been reported that SiO$_2$ exhibit similar characteristics [37].

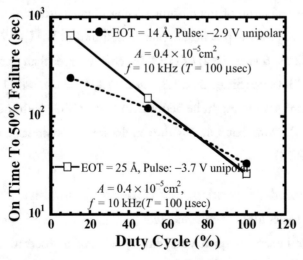

FIGURE 2.16: A larger lifetime enhancement is observed as thickness increases.

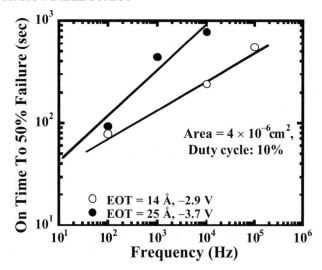

FIGURE 2.17: As frequency increases, thicker HfO$_2$ has more accelerated lifetime enhancement.

Also, as thickness decreases, steeper voltage acceleration factors for both dc and unipolar stressing were observed (Fig. 2.18) as also observed in SiO$_2$. In fact, this indicates that there are similarities in lots of breakdown behaviors.

Stress-induced leakage current (SILC) has been widely used as a measure of the defect generation rate for ultrathin gate oxide. The relatively small increase

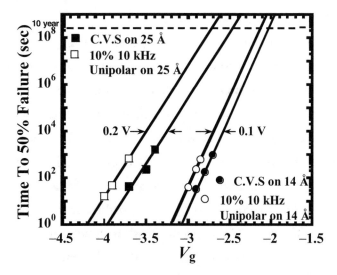

FIGURE 2.18: The thickness dependence of lifetime enhancement is observed.

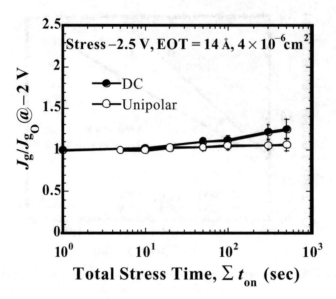

FIGURE 2.19: Stress-induced leakage current characteristics of 1.4 nm HfO$_2$ under dc and unipolar stressing.

in SILC for HfO$_2$ was used to explain the low defect generation and the charge trapping in this experiment. Fig. 2.19 shows the SILC characteristics of 1.4-nm HfO$_2$ under dc and unipolar stressing. As shown in this figure, unipolar stressing exhibits less SILC characteristics than does dc stressing.

In general, lower charge trapping in dynamic stressing was observed compared to constant voltage stressing. The dynamics of the charge trapped near the anode interface in unipolar stressing mode for SiO$_2$ is so slow ($\tau_{in} \approx 0.1$ s) compared to the stress period ($t_{on} = 10~\mu$s) that there is a significant reduction of the effective trapped charge [36]. One possible explanation is that time (τ_{in}) for charge to be trapped in HfO$_2$ is longer than t_{on} of high-frequency unipolar stress, resulting in less charge trapping under unipolar stressing as well as detrapping during the off time. It is also thought that the time (τ_{in}) may have different values as a function of thickness. For HfO$_2$ with EOT $= 2.5$ nm, as stress time increases, the difference between dc and unipolar stressing in terms of SILC characteristics becomes larger than that of HfO$_2$ with EOT $= 14$ Å (Fig. 2.20).

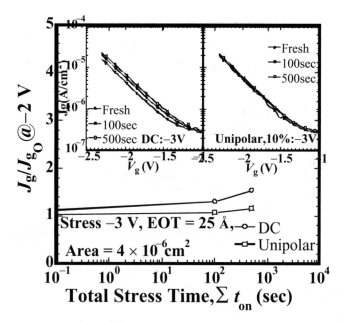

FIGURE 2.20: Stress-induced leakage current characteristics of 2.5 nm HfO$_2$ under dc and unipolar stressing.

The increased amount of SILC for thick HfO$_2$ (25 Å) is approximately twice as much as that of thin HfO$_2$ (14 Å). These SILC behaviors support lower enhancement of ac stress lifetime projection for thinner dielectrics in comparison to thick HfO$_2$. C–V hysteresis characteristics after ac stressing shows less degradation compared to those after dc stressing (Fig. 2.21). Hysteresis is defined as ΔV_{fb} on the C–V curves. H_0 represents hysteresis for a fresh device. It is thought that less charge trapping related to low defect generation under ac stressing may result in less shifting the flat band voltage corresponding to hysteresis.

2.4.4 Charge Trapping Characteristics During On and Off Time

It is believed that the longer time-to-breakdown of HfO$_2$ under unipolar stress is attributed to charge detrapping during t_{off} and reduced trapping during t_{on} [38]. The experimental results indicated that the dynamic reliability strongly depends on the values of on and off time. It is shown that there is a critical time for enhancement

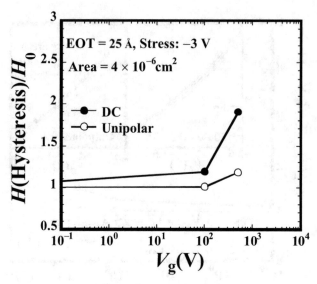

FIGURE 2.21: Hysteresis characteristics of 2.5 nm HfO$_2$ under dc and unipolar stressing.

in charge detrapping and reduction in charge trapping. Fig. 2.22 shows the increase of t_{BD} with unipolar ac stress over dc stressing as a function of duty cycle. As on time (t_{on}) increases, the time-to-breakdown approaches to the dc t_{BD} value. To further understand this behavior, observe the figure where it is quantitatively shown that when t_{on} is less than about 10^{-3} sec, the characteristic time for charge to be trapped

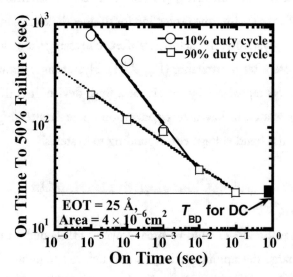

FIGURE 2.22: As on time (t_{on}) increases, time-to-breakdown approaches to the dc t_{BD} value.

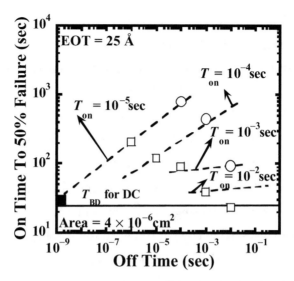

FIGURE 2.23: When t_{on} is less than about 10^{-3} sec, the detrapping is more effective.

in HfO$_2$ (τ_{in}) is longer than t_{on} of unipolar stress. This results in less charge trapping and thus the detrapping during the off period becomes significantly more effective (Figs. 2.22 and 2.23) [39].

2.5   CONCLUSION

We have shown that the Weibull slopes of soft and hard breakdown for HfO$_2$ (EOT = 14 Å) gate dielectrics were found to be $\beta = 1.4$ and $\beta = 2$, respectively. In order to obtain accurate β value, various sizes of capacitors and different stress voltages were used. Area dependence of gate scaling has been proven using obtained β value. For HfO$_2$ with EOT = 1.4 nm, the soft breakdown predominantly occurs as a first breakdown event. For practical scaling purposes, soft breakdown behavior might be one of scaling issues of high-k dielectrics because it shows worse reliability characteristics.

Thickness dependence of breakdown behavior has been observed and analyzed for HfO$_2$. The dependence seems to follow the percolation model. Weibull slope ($\beta \approx 4.0$) of thick HfO$_2$ (EOT = 2.5 nm) is smaller than that of SiO$_2$

($\beta \approx 8.0$). On the other hand, thin HfO$_2$ (EOT $= 1.4$ nm) exhibits similar β ($\sim$2), which is still slightly smaller than that of SiO$_2$ ($\beta = 2.2$–2.5). Low β for thick HfO$_2$ would result in reduction of the lifetime scaling to percentile and total oxide area on chip. However, for practical application, ultrathin HfO$_2$ with EOT < 2 nm be used. Thus, β value of HfO$_2$ is actually larger than that of SiO$_2$ with the same EOT value (SiO$_2$: $\beta = 1.3$–1.4 for EOT $= 1.5$ nm) and therefore lifetime scaling might be better than SiO$_2$ when compared at the same EOT value.

The dielectric breakdown of HfO$_2$ under unipolar stressing has been analyzed in comparison to dc stressing. Longer lifetime of HfO$_2$ under dynamic ac stressing has been observed in comparison to constant voltage stress. Higher frequency, lower duty cycle, and thicker dielectrics result in larger lifetime enhancement ($t_{BD,ac}/t_{BD,dc}$). Stress-induced leakage current and hysteresis degradation were also found to be less severe under ac stress. The improved reliability under ac stress is believed to be due to the combination of reduced charge trapping during on-time and charge detrapping during off-time as compared to dc stressing. In fact, it turns out that when t_{on} is less than about 10^{-3} sec, the characteristic time for charge to be trapped in HfO$_2$(τ_{in}) is longer than t_{on} of unipolar stress, resulting in less charge trapping, and the detrapping becomes significantly more effective.

REFERENCE

[1] R. H. Dennard, F. H. Gaensslen, H. N. Yu, V. L. Rideout, E. Bassous, and A. R. LeBlanc, "Design of ion-implanted MOSFETs with very small physical dimensions," *IEEE J. Solid-State Circuits*, Vol. SC-9, p. 256, 1974. doi:10.1109/JSSC.1974.1050511

[2] S.-H. Lo, D. A. Buchanan, Y. Taur, and W. Wang, "Quantum-mechanical modeling of electron tunneling current from the inversion layer of ultra-thin-oxide nMOSFET's," *IEEE Trans. Electron Devices*, Vol. ED-18, No. 5, pp. 209–211, 1997.

[3] T.-S. Chen, D. Hadad, V. Valu, V. Jiang, S.-H. Kuah, P. C. McIntyre, S. R. Summerfelt, J. M. Anthony, and J. C. Lee, "Ir-electroded BST thin film

capacitors for 1 giga-bit DRAM application," in *IEEE Int. Electron Devices Meeting*, 1996, pp. 679–682.

[4] D. Barlage, R. Arghavani, G. Deway, M. Doczy, B. Doyle, J. Kavalieros, A. Murthy, B. Roberds, P. Stokley, and R. Chau, "High-frequency response of 100 nm integrated CMOS transistors with high-k gate dielectrics," in *IEEE Int. Electron Devices Meeting*, 2001, pp. 231–234.

[5] M. Balog, M. Schieber, M. Michman, and S. Patai, "Chemical vapor deposition and characterization of HfO_2 films from organo-hafnium compounds," *Thin Solid Films*, Vol. 41, pp. 247–259, 1997. doi:10.1016/0040-6090(77)90312-1

[6] B. Cheng, M. Cao, R. Rao, A. Inani, P. V. Voorde, W. M. Greene, J. M. C. Stork, Z. Yu, P. M. Zeitzoff, and J. C. S. Woo, "The impact of high-k gate dielectrics and metal gate electrodes on sub-100nm MOSFET's," *IEEE Trans. Electron Devices*, Vol. 46, pp. 1537–1544, 1999. doi:10.1109/16.772508

[7] A. Kumar, T. H. Ning, M. V. Fischetti, and E. Gusev, "Hot-carrier charge trapping and reliability in high-k dielectrics," *VLSI Symp. Tech. Dig.*, pp. 152–153, 2002.

[8] M. Koyama, K. Suguro, M. Yoshiki, Y. Kamimuta, M. Koike, M. Ohse, C. Hongo, and A. Nishiyama, "Thermally stable ultra-thin nitrogen incorporated ZrO_2 gate dielectric prepared by low temperature oxidation of ZrN," *Tech. Dig. IEDM*, pp. 459–462, 2001.

[9] C. H. Lee, H. F. Luan, W. P. Bai, S. J. Lee, T. S. Jeon, Y. Senzaki, D. Roberts, and D. L. Kwong, "MOS characteristics of ultra thin rapid thermal CVD ZrO_2 and Zr silicate gate dielectrics," *Tech. Dig. IEDM*, pp. 27–30, 2000.

[10] L. Kang, K. Onishi, Y. Jeon, B. Lee, C. Kang, W. Qi, R. Nieh, S. Gopalan, R. Choi, and Jack C. Lee, "MOSFET devices with polysilicon on single-layer HfO_2 high-k dielectrics," *Tech. Dig. IEDM*, p. 35–38, 2000.

[11] B. H. Lee, R. Choi, L. G. Kang, S. Gopalan, R. Nieh, K. Onishi, Y. J. Jeon, W. J. Qi, C. S Kang, and J. C. Lee, "Characteristics of TaN gate MOSFET with ultrathin hafnium oxide (8 Å–12 Å)," *Tech. Dig. IEDM.*, p. 39, 2000.

[12] J. Robertson, "Band offsets of wide-band-gap oxides and implications for future electronic devices," *J. Vacuum Sci. Tech. B*, Vol. 18, pp. 1785–1791, 2000. doi:10.1116/1.591472

[13] R. Choi, K. Onishi, C. S. Kang, S. Gopalan, R. Nieh, Y. H. Kim, J. H. Han, S. Krishnan, H. J. Cho, A. Sharriar, and J. C. Lee, "Fabrication of high quality ultra-thin HfO$_2$ gate dielectric MOSFETs using deuterium anneal," *Tech. Dig. IEDM*, pp. 613–616, 2002.

[14] K. Onishi, R. Choi, C. S. Kang, H. J. Cho, S. Gopalan, R. Nieh, S. Krishnan, and Jack C. Lee, "Effects of high-temperature forming gas anneal on HfO$_2$ MOSFET performance," *Tech. Dig. Int. Symp. VLSI.*, pp. 22–23, 2002.

[15] H. R. Huff, A. Hou, C. Lim, Y. Kim, J. Barnett, G. Bersuker, G. A. Brown, C. D. Young, P. M. Zeitzoff, J. Gutt, P. Lysaght, M. I. Gardner, and R. W. Murto, "Integration of high-*k* gate stacks into planar, scaled CMOS integrated circuits," in *Conf. on Nano and Giga Challenges in Microelectronics*, 2002, pp. 1–18.

[16] R. Degraeve, G. Groeseneken, R. Bellens, M. Depas, and H. E. Maes, "A consistent model for the thickness dependence of intrinsic breakdown in ultra-thin oxides," *Tech. Dig. IEDM*, pp. 863–866, 1995.

[17] M. Houssa, T. Nigam, P. W. Mertens, and M. M. Heyns "Model for the current-voltage characteristics of ultra thin gate oxide after soft breakdown," *J. App. Phys.*, Vol. 84, pp. 4351–4355, 1998. doi:10.1063/1.368654

[18] J. H. Stathis and D. J. DiMaria, "Reliability projection for ultra-thin oxides at low voltage," *Tech. Dig. IEDM*, pp. 167–170, 1998.

[19] J. H. Stathis, "Physical and predictive models of ultra thin oxide reliability in CMOS devices and circuits," in *IEEE Reliability Physics Symp.*, 2001, pp. 132–149.

[20] T. Nigam, R. Degraeve, G. Groeseneken, M. M. Heyns, and H. E. Maes, "Constant current charge-to-breakdown: Still a valid tool to study the reliability of MOS structures?," in *IEEE Reliability Physics Symp.*, 1998, pp. 62–69.

[21]   F. Monsieur, E. Vincent, D. Roy, S. Bruyere, J. C. Vildeuil, G. Pananakakis, and G. Ghibaudo, "A thorough investigation of progressive breakdown in ultra-thin oxides. Physical understanding and application for industrial reliability assessment," in *IEEE Reliability Physics Symp.*, 2002, pp. 45–54.

[22]   J. Sune, E. Y. Wu, D. Jimbez', R. P. Vollertsen, and E. Miranda, "Understanding soft and hard breakdown statistics, prevalence ratios and energy dissipation during breakdown runaway," *Tech. Dig. IEDM*, pp. 117–120, 2001.

[23]   Jordi Suñé and Ernest Wu, "Statistics of successive breakdown events for ultra-thin gate oxides," *Tech. Dig. IEDM*, pp. 147–150, 2002.

[24]   R. Degraeve, B. Kaczer, M. Houssa, G. Groeseneken, M. Heyns, J. S. Jeon, and A. Halliyal, "Analysis of high voltage TDDB measurements on Ta_2O_5/SiO_2 stack," *Tech. Dig. IEDM*, pp. 327–330, 1999.

[25]   P. D. Kirsch, C. S. Kangl, J. Lozano, J. C. Lee, and J. G. Ekerdt, "Electrical and spectroscopic comparison of HfO_2/Si interfaces on nitrided and un-nitrided Si (100)," *J. Appl. Phys.*, Vol. 91, pp. 4353–4363, 2002. doi:10.1063/1.1455155

[26]   David K. Cheng, *Fundamentals of Engineering Electromagnetics*, Addison-Wesley, Reading, MA, p. 114, 1993.

[27]   Y.-H. Kim, K. Onishi, C. S. Kang, H.-J. Choi, R. Nieh, S. Gopalan, R. Choi, J. Han, S. Krishnan, and J. C. Lee, "Area dependence of TDDB characteristics for HfO_2 gate dielectrics," *IEEE Elec. Dev. Lett.*, Vol. 23, pp. 594–596, 2002. doi:10.1109/LED.2002.803751

[28]   T. Kauerauf, R. Degraeve, E. Cartier, C. Soens, and G. Groesenenken, "Low Weibull slope of breakdown distributions in high-*k* layers," *IEEE Elec. Dev. Lett.*, Vol. 23, pp. 215–217, 2002. doi:10.1109/55.992843

[29]   A. Kerber, E. Cartier, R. Degraeve, L. Pantisano, Ph. Roussel, and G. Groeseneken, "Strong correlation between dielectric reliability and charge trapping in SiO_2/Al_2O_3 gate stacks with TiN electrodes," *Tech. Dig. Symp. VLSI*, pp. 76–77, 2002.

[30]   T. Kauerauf, R. Degraeve, E. Cartier, B. Govoreanu, P. Blomme, B. Kaczer, L. Pantisano, A. Kerber, and G. Groeseneken, "Towards understanding

degradation and breakdown of SiO$_2$/high-k stacks," *Tech. Dig. IEDM*, pp. 521–524, 2002.

[31] D. J. DiMaria and J. H. Stathis, "Explanation for the oxide thickness dependence of breakdown characteristics of metal-oxide-semiconductor structures," *Appl. Phys. Lett.*, Vol. 70, No. 20, pp. 2708–2710, 1997. doi:10.1063/1.118999

[32] J. H. Stathis, "Percolation models for gate oxide breakdown," *J. Appl. Phys.*, Vol. 86, No. 10, pp. 5757–5766, 1999. doi:10.1063/1.371590

[33] Y.-H. Kim, K. Onishi, C. S. Kang, H.-J. Cho, R. Choi, S. Krishnan, Md. S. Akbar, and J. C. Lee, "Thickness dependence of Weibull slopes of HfO$_2$ gate dielectrics," *IEEE Elect. Device Lett.*, pp. 40–42, 2003.

[34] Y. Fong, I. C. Chen, S. Holland, J. Lee and C. Hu, "Dynamic stressing of thin oxides," *Tech. Dig. IEDM.*, p. 863, 1995.

[35] E. Rosenbaum, Z. Liu, and C. Hu, "Silicon dioxide breakdown lifetime enhancement under bipolar bias conditions," *IEEE Trans. Elec. Dev.*, Vol. 40, p. 2287, 1993. doi:10.1109/16.249477

[36] G. Ghidini, D. Brazzelli, C. Clementi, and F. Pellizzer, "Charge trapping mechanism under dynamic stress and its effect on failure time," *Proc. of IEEE Reliability Physics Symp.*, 1999, p. 88.

[37] P. Chaparala, J. S. Suehle, C. Messick, and M. Roush, "Electrical field dependent dielectric breakdown of intrinsic SiO$_2$ films under dynamic stress," *Proc. of IEEE Reliability Physics Symp.*, 1996, p. 61. doi:full_text

[38] Y. H. Kim, K. Onishi, C. S. Kang, R. Choi, H.-J. Cho, R. Nieh, J. Han, S. Krishnan, A. Shahriar, and J. C. Lee, "Hard and soft-breakdown characteristics of ultra-thin HfO$_2$ under dynamic and constant voltage stress," *Tech. Dig. IEDM.*, p. 629, 2002.

[39] Y. H. Kim, K. Onishi, C. S. Kang, R. Choi, H.-J. Cho, S. Krishnan, A. Shahriar, and J. C. Lee, "Dynamic reliability characteristics of ultra-thin HfO$_2$," in *IEEE Reliability Physics Symp.*, 2003, pp. 46–50.

C H A P T E R   3

Impact of High Temperature Forming Gas and D$_2$ Anneal on Reliability of HfO$_2$ Gate Dielectric

3.1 PREVIOUS RESULTS

Recently, high-k dielectrics have attracted a great deal of attention as the replacement of gate oxides because they meet the stringent requirements for leakage current and equivalent oxide thickness (EOT). Among various high-k gate dielectric materials, HfO$_2$ has been an attractive material because it has demonstrated good device characteristics and is compatible with the polysilicon gate process [1, 2]. However, according to previous reports [3, 4] MOSFET performance of HfO$_2$ devices has showed in part degraded mobility characteristics in comparison to that of SiO$_2$ devices. Improving the quality of the interface between the high-k dielectric and the Si substrate has been one of the major issues in the high-k gate dielectrics. It has also been reported that the degraded mobility of HfO$_2$ MOSFET could be improved by high temperature forming gas and deuterium gas annealing [5, 6]. In other word, high temperature forming gas or deuterium anneal improved interface state density leading to high drain current and carrier mobility. These results, as might be expected, indicate that the high-*k* device performance is significantly affected by interface quality. However, it is not fully clear as how much device performance improved by various optimization of interface structures and processing (e.g. anneal condition). In this section, we will discuss the carrier mobility in HfO_2 MOSFET's with regard to different surface preparations and reliability characteristics as D_2 anneal temperature increases.

3.2   EFFECT OF D_2 ANNEAL ON VARIOUS SURFACE PREPARATIONS

3.2.1   Previous Work of Effect of D_2 Anneal on Various Surface Preparations

In this section, we are focusing on the interface engineering and the improvement obtained by deuterium annealing. It not only confirms previous results and but also finds dependency of anneal effects in terms of interface layer preparation.

3.2.2   Predeposition Treatment and High-*k* Deposition

Three different predeposition surface treatments were prepared: HF last, O_3 clean, and HF/NH_3 700°C, 15 sec. This was followed by atomic layer deposition (ALD) HfO_2 and CVD Hf silicate deposition. It has been found that O_3 clean led to more distinct interface and smoother high-*k* film subsequently deposited by ALD method [7]. NH_3 anneal has been found especially advantageous in minimizing the subsequent growth of the interface oxide-like film either during predeposition anneal or postdeposition anneal in an oxygen ambient with regards to EOT and leakage current [8]. After high-*k* layer deposition, postdeposition anneals (PDA) were performed. N_2 600°C and NH_3 700°C were applied for both ALD HfO_2 and CVD Hf silicate, respectively. In any case, PDA reduces leakage current by

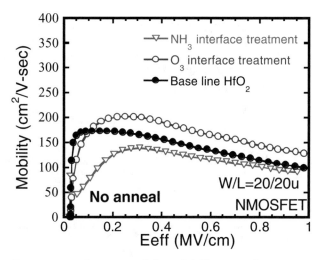

FIGURE 3.1: Comparison of carrier mobility of different surface treatments. O$_3$-treated device has higher mobility than others.

densification of the dielectrics. LPCVD amorphous silicon gate electrode was deposited at 560°C/2 h in SiH$_4$ plus H$_2$ and annealed at 1000°C. Deuterium gas anneal was performed at 550°C for 30 min after metallization.

3.2.3   Result and Discussion

Fig. 3.1 shows the comparison of the mobility of three different HfO$_2$ interface layers before deuterium annealing. All three devices have a similar EOT value ($\approx$14.4 Å). Mobility of the devices was extracted from the inversion capacitance on the basis of a split CV measurement [9]. It is known that O$_3$ treated interface layer has a low interface state density [7] (D_{it} value $\sim$10^{10} eV/cm^2), which enhances carrier mobility in the channel and a good threshold swing value as well. Because of the high nitrogen..., NH$_3$ treated interface layer suffers high interface state density and fixed charge leading to charge scattering. This interface quality may significantly impact mobility characteristics as shown in Fig. 3.1.

In general, in order to scale down HfO$_2$, it is indispensable to have a good diffusion barrier against further oxidation at the interface as well as have a high

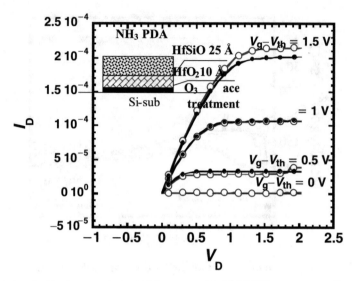

FIGURE 3.2: I_d–V_d characteristics of O_3-treated MOSFET. D_2 anneal did not change drain current.

dielectric constant. O_3-treated interface layer has good interface quality but it may not be scalable because it has weak diffusion barrier properties and low dielectric constant as well. On the other hand, NH_3 treatment at the interface may be more advantageous in terms of scaling because of its barrier property and high dielectric constant. Therefore, there might be trade off between the scaling and device performances. In previous reports [5, 6] it was reported that forming gas and deuterium gas anneal have improved HfO_2 MOSFET performances. It is worth questioning that these anneals are effective for all interface layers. Fig. 3.2 shows the deuterium annealing performed for O_3-treated HfO_2 devices. For an accurate evaluation, exactly the same device was measured before and after D_2 anneal. As can be seen, deuterium annealing does not improve the drain current of HfO_2 devices, while the NH_3-treated interface layer of HfO_2 devices shows a significant improvement in drain current after D_2 anneal (Fig. 3.3).

Fig. 3.4 shows the moderate improvement in control HfO_2. Amongst all surface treatments, NH_3-treated interface layer showed the most significant improvement by D_2 anneal.

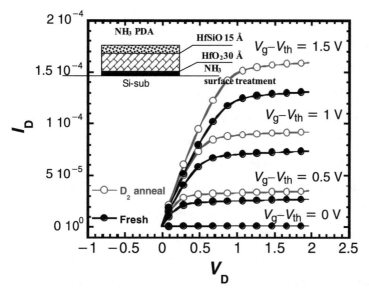

FIGURE 3.3: Drain current of NH_3-treated HfO_2 MOSFET shows significant improvement.

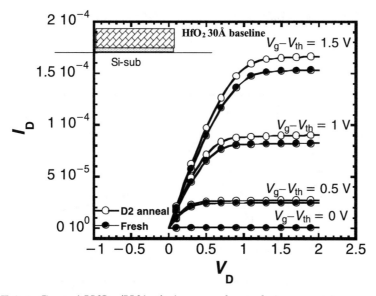

FIGURE 3.4: Control HfO_2 (Hf last) shows moderate drain current improvement but lower than that of NH_3 treatment.

It would have been more effective if D_2 anneal was done prior to metallization because this would have allowed higher temperature. Furthermore, it is worth noting that short channel devices will have more improvement because of shorter diffusion length of D_2. An interface layer of control HfO_2 is known to grow due to subsequent thermal processing (e.g. PDA). This interface quality is not so good as that of O_3-treated devices, but is better than that of NH_3. This may be the reason why the improvement of base line HfO_2 by D_2 anneal is more noticeable than that of O_3, but is smaller than that of NH_3. For the scaling purpose, NH_3 surface treatment may be required to suppress oxygen diffusion and increase dielectric constant. Therefore, it is suggested that D_2 annealing may be a possible solution for NH_3 surface treatment to achieve compatible performance in comparison to O_3 treatment, while keeping its merits. In addition to a performance characterization, a reliability study of those three samples is being investigated to see whether D_2 anneal has a selective impact on NH_3, O_3, and HF last surface treatment.

Fig. 3.5 shows mobility improvement of NH_3-treated device before and after D_2 anneal.

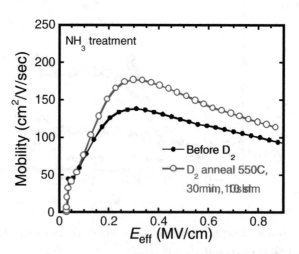

FIGURE 3.5: Mobility improvement of NH_3-treated device before and after D_2 anneal.

3.2.4 Conclusion

Three different surface treated interface layers of HfO_2 were investigated. Mobility of HfO_2 device was strongly dependent on the interface treatments. O_3 surface treatment showed higher mobility than those of HF last and NH_3. This might result from low interface state density of O_3-treated device as good as SiO_2. Deuterium gas anneal was performed on three different surface treated interface layers. It was found that selective drain current improvements were observed. NH_3 surface treatment showed most improved drain current after D_2 anneal, while O_3 treatment showed a negligible increase in drain current. In other words, NH_3 surface treatment that has a poor interface quality showed the most improvement in drain current. The good quality of O_3-treated interface did not show significant drain current increases. However, the reliability study needs to be investigated because a high electrical stress may have different results.

3.3 EFFECTS OF HIGH TEMPERATURE FORMING GAS IN TERMS OF RELAIBILITY

In this section, the effects of forming gas anneal temperature and dielectric leakage current on HfO_2 devices are discussed in terms of Weibull slope and breakdown distribution. These effects result from higher hydrogen induced defect generation rate under high electric field for the high temperature annealed devices. In addition to forming gas anneal, we show that dielectric leakage current influences device lifetime area scaling value. The relation between leakage current and device breakdown characteristics will be discussed in this section. Onishi *et al.* have shown that forming gas anneal improves interface quality of HfO_2 by lowering interface state density, but degradation of breakdown characteristics was also implied for high temperature forming gas anneal [10, 11]. Choi *et al.* have reported that deuterium anneal improves HfO_2 interface quality without significant degradation of breakdown characteristics [12]. However, a more detailed study of the reliability characteristics in terms of the forming gas anneal temperature and the dependence on the dielectric leakage characteristics is needed. These are the focus of the section.

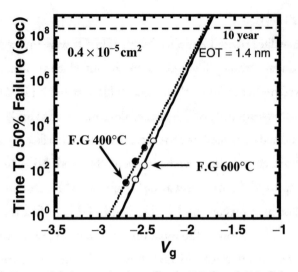

FIGURE 3.6: A 10 years lifetime projection of both 400°C and 600°C forming gas anneals.

3.3.1   Reliability Characteristics: TDDB, Weibull Slope, and Lifetime Projection

As shown in Fig. 3.6, at 400°C and 600°C forming gas anneal (30 min) did not show significant difference in terms of TDDB lifetime projection. However, as indicated in previous reports [10, 11], time-zero dielectric breakdown behavior of HfO$_2$ changed after high temperature forming gas anneals.

The comparison of TZDB characteristics of various device with low and high temperature forming gas anneals sizes is shown in Fig. 3.7. As can be seen, TZDB of low temperature (400°C) forming gas annealed samples shows tighter distribution in comparison to that of high temperature forming gas annealed samples. A plausible mechanism can be related to the higher hydrogen induced defect generation rate under high electric field for the high temperature annealed devices. For practical operating voltage (low field), on the contrary, high temperature forming gas anneal significantly improves MOSFET performances [10, 11].

Fig. 3.8 shows the comparison of Weibull slopes (β) of both 400°C and 600°C forming gas annealed samples. The lower β value for the higher temperature forming gas annealed samples would result in a reduction of lifetime scaling to percentile and total oxide area on chip. Our results also indicate that the gate area

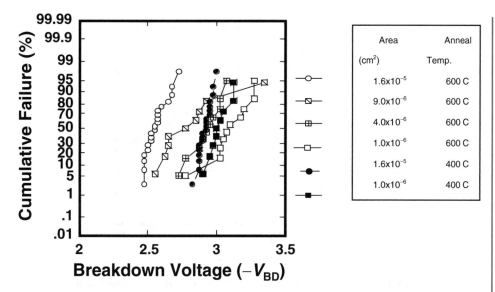

FIGURE 3.7: Time-zero dielectric breakdown distribution of 400°C and 600°C forming gas annealed samples. The higher the anneal temperature, the wider the distribution.

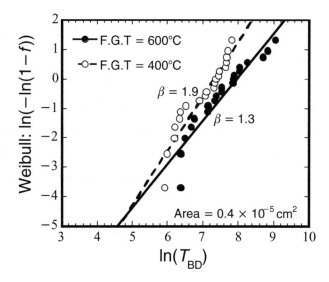

FIGURE 3.8: Weibull slope comparison between high and low temperature forming gas anneal.

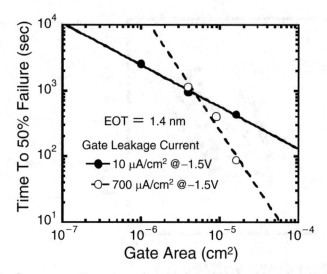

FIGURE 3.9: Gate area scaling comparison between high and low gate leakage current.

scaling is strongly dependent on gate leakage currents (Fig. 3.9). A worse scaling factor may result from a high defect density, which leads to high leakage current.

3.4   CONCLUSION

High temperature forming gas anneal shows little degradation in terms of TDDB lifetime projection, whereas it showed good interface characteristics. This is due to hydrogen-related defect generation for high electric field stress. It is also worth mentioning that area scaling is very sensitive to gate leakage current. In other words, initial defect density may be critical to evaluate high-k dielectrics, and need to be taken into consideration. Also area scaling and TDDB results as well as Weibull slope of high temperature forming gas annealed samples show smaller β value than that of 400°C sample. In summary, high temperature forming gas anneal shows good interface quality of device and negligible degradations were observed.

REFERENCE

[1]   L. Kang, K. Onishi, Y. Jeon, B. Lee, C. Kang, W. Qi, R. Nieh, S. Gopalan, R. Choi, and J. C. Lee, "MOSFET devices with polysilicon on single-layer HfO$_2$ high-k dielectrics," *Tech. Dig. IEDM*, p. 35–38, 2000.

[2]   K. Onishi, C. Kang, R. Choi, H. Cho, S. Gopalan, R. Nieh, E. Dharmarajan, and J. C. Lee, "Reliability characteristics, including NBTI, of polysilicon gate HfO$_2$ MOSFET's," *Tech. Dig. IEDM,* pp. 659–662, 2001.

[3]   S. J. Lee, H. F. Luan, C. H. Lee, T. S. Jeon, W. P. Bai, Y. Senzaki, D. Roberts, and D. L. Kwong "Performance and reliability of ultra thin CVD HfO$_2$ gate dielectrics with dual poly-Si gate electrodes," *Tech. Dig. Int. Symp. VLSI.,* pp. 133–134, 2002.

[4]   C.-H. Lee, J. J. Lee, W. P. Bai, S. H. Bae, J. H. Sim, X, Lei, R. D. Clark, Y. Harada, M. Niwa, and D. L. Kwong, "Self-aligned ultra thin HfO$_2$/CMOS transistors with high quality CVD TaN gate electrode," *Tech. Dig. Int. Symp. VLSI.,* pp. 82–83, 2002.

[5]   K. Onishi, R. Choi, C. S. Kang, H. J. Cho, S. Gopalan, R. Nieh, S. Krishnan, and Jack C. Lee, "Effects of high-temperature forming gas anneal on HfO$_2$ MOSFET performance," *Tech. Dig. Int. Symp. VLSI.,* pp. 22–23, 2002.

[6]   R. Choi, K. Onishi, C. S. Kang, S. Gopalan, R. Nieh, Y. H. Kim, J. H. Han, S. Krishnan, H. J. Cho, A. Sharriar, and J. C. Lee, "Fabrication of high quality ultra-thin HfO$_2$ gate dielectric MOSFETs using deuterium anneal," *Tech. Dig. IEDM,* pp. 613–616, 2002.

[7]   H. R. Huff, A. Hou, C. Lim, Y. Kim, J. Barnett, G. Bersuker, G. A. Brown, C. D. Young, P. M. Zeitzoff, J. Gutt, P. Lysaght, M. I. Gardner, and R. W. Murto, "Integration of high-*k* gate stacks into planar, scaled CMOS integrated circuits," in *Conf. Nano and Giga Challenges in Microelectronics* 2002, pp. 1–18.

[8]   R. Choi, C. S. Kang, B. H. Lee, K. Onishi, R. Nieh, S. Gopalan, E. Dharmarajan, and J. C. Lee, "High-quality ultra-thin HfO$_2$ gate dielectric MOSFETs with TaN electrode and nitridation surface preperation," *Tech. Dig. Int. Symp. VLSI.,* pp. 15–16, 2001.

[9]   J. Koomen, "Investigation of the MOST channel conductance in weak inversion," *Solid-State Electron.,* Vol. 16, pp. 801–810, 1973. doi:10.1016/0038-1101(73)90177-9

[10] K. Onishi, C. Kang, R. Choi, H. Cho, S. Gopalan, R. Nieh, E. Dharmarajan, and Jack C. Lee, "Reliability characteristics, including NBTI, of polysilicon gate HfO_2 MOSFET's," *Tech. Dig. IEDM*, pp. 659–662, 2001.

[11] K. Onishi, C. S. Kang, R. Choi, H. J. Cho, S. Gopalan, R. Nieh, S. Krishnan, and Jack C. Lee, "Improvement of surface carrier mobility of HfO2 MOSFET's by high-temperature forming gas annealing," in *IEEE Trans. Electron Devices*, Vol. 50, No. 2, p. 384–390, February 2003. doi:10.1109/TED.2002.807447

[12] R. Choi, K. Onishi, C. S. Kang, S. Gopalan, R. Nieh, Y. H. Kim, J. H. Han, S. Krishnan, H. J. Cho, A. Sharriar, and J. C. Lee, "Fabrication of high quality ultra-thin HfO_2 gate dielectric MOSFETs using deuterium anneal," *Tech. Dig. IEDM*, pp. 613–616, 2002.

C H A P T E R 4

Effect of Barrier Height and the Nature of Bilayer Structure of HfO$_2$ with Dual Metal Gate Technology

4.1 MOTIVATION

High-k dielectrics, HfO$_2$, ZrO$_2$, and Al$_2$O$_3$, have been investigated to replace SiO$_2$ in order to reduce high leakage current in ultrathin regime [1–20]. High-k dielectrics are especially advantageous for low-power application and for thickness uniformity control.

High-k dielectrics have been known to be "trap-rich" materials. A number of efforts have been carried out to improve high-k dielectrics characteristics. For example, high temperature deuterium and forming gas anneals and O$_3$ surface treatments have been shown to reduce interface state [21–23]. The origin of traps in high-k dielectrics, however, still remains a question. These preexisting traps may play an important role in dielectric wear-out as well as device performance. It has been reported that Weibull slope of HfO$_2$ is smaller than that of SiO$_2$ with the same physical thickness [24]. Possible explanations for this might be either smaller critical defect densities due to preexisting traps, extrinsic defects, or bimodal defect generations [24–26]. Besides, metal gate electrodes are expected to be needed in the near future. However, there has been a lack of the experimental evidences and study on the gate electrode effects on wear-out characteristics of high-k devices. In this chapter we study the wear-out characteristics and Weibull slope behaviors of HfO$_2$ devices in terms of interface layer, high-k layer, and barrier height by gate electrode workfunction.

In this chapter, we present the effect of gate electrodes and barrier heights on the breakdown characteristics and Weibull slopes of HfO$_2$ MOS devices. Higher Weibull slope (β) of Ru gate electrode has been observed when compared with that of Ru–Ta alloy. The higher β in Ru devices is due to smaller charge fluence, which results from relatively higher barrier height. Varying SiO$_2$ and HfO$_2$ were investigated in order to understand ultrathin HfO$_2$ gate stack structure. We found that there is bimodal defect generation rate on high-k/SiO$_2$ stack. Two-step breakdown process was clearly observed and Weibull slope of soft breakdown (first breakdown) shows lower β value compared to that of hard breakdown (second breakdown). Soft breakdown characteristics were dependent on the barrier heights. The bimodal defect generations are believed to be due to the breakdown in interface and bulk layer. Weibull slope of high-k gate stacks is a strong function of the interface layer thickness and the barrier height by gate electrode workfunction, whereas it is a weak function of high-k layer thickness.

4.2   EXPERIMENTAL PROCEDURE

HfO$_2$ was deposited using reactive dc magnetron sputtering with O$_2$ modulation technique [27], followed by postdeposition annealing (PDA) at 500°C. Ru and Ru–Ta gate electrodes were deposited on top of HfO$_2$. W was deposited on the sample with Ru–Ta gate. The electrodes were patterned using lift-off process, followed by forming gas annealing at 400°C for 30 min. To keep the samples from possible metal diffusion into dielectrics, none of the high temperature processing was performed after gate electrode formation.

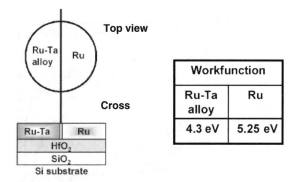

FIGURE 4.1: Metal gates deposited on the same dielectric structure.

4.3   RESULTS AND DISCUSSION

Fig. 4.1 shows device preparations and workfunctions of Ru–Ta alloy and Ru gate electrode.

It has been reported that Ru–Ta alloy gate electrode is NMOS compatible, while Ru is PMOS compatible [28]. According to previous results on Ru–Ta alloy, workfunction of the alloy is very sensitive to thermal budget and need to be controlled precisely. When one integrates metal gate with gate dielectrics, one of challenging issues is metallic ion diffusion. To prevent gate dielectric from metallic diffusion, thermal budget was controlled below 600°C, and a stable workfunction could be achieved with good dielectric properties. Workfunction difference between Ru and Ru–Ta alloy is ~1 eV. The Ru and Ru–Ta alloy gate electrode were deposited on the same dielectric stacks and same type of substrate. Therefore, we could compare only the dielectric property with different workfunction materials. It is worth noting that, for the first time, we carried out detailed reliability experiments on exactly same quality of high-k gate dielectrics with dual metal gate electrode application.

For gate injection, relatively slow breakdown wear-out was observed including SILC and soft breakdown (Figs. 4.2 and 4.3).

On the other hand, fast immediate breakdown immediately after soft breakdown was observed for substrate injection with thick interface layer (Figs. 4.4 and 4.5).

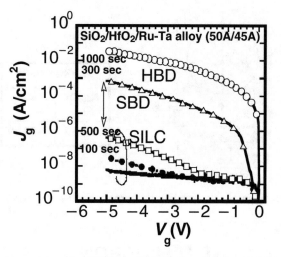

FIGURE 4.2: Slow transition from SBD to HBD observed for NMOS capacitor.

HfO$_2$ gate stack undergoes various tunneling conductions, and defect generations under the different polarities. Thus, the role that the interfacial layer plays in triggering soft breakdown can be higher defect generation rate due to high stress field and intense charge fluence through the interface layer under substrate injection. In case of thicker interface layer under substrate injection, it would result in immediate hard breakdown immediately after soft breakdown because the voltage

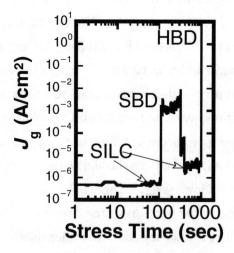

FIGURE 4.3: Current fluctuation observed for thick high-*k* stack.

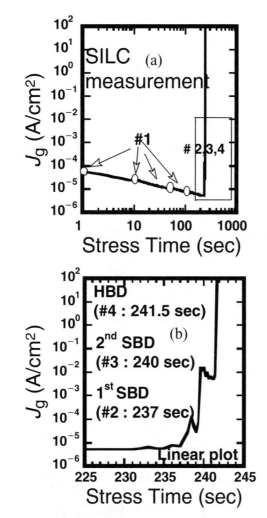

FIGURE 4.4: (a) Stress time vs. leakage current for substrate injection. (b) Magnified linear plot shows soft and hard breakdown process for given short period.

drop across an interface layer increases the voltage across high-k layer after interface layer breakdown (Figs. 4.4 and 4.5). In other words, if an interface layer thickness is controlled within 5–9 Å with fairly thick high-k layer (>45 Å), it will show the slow transition to hard breakdown after soft breakdown (interface layer breakdown) under substrate injection because high-k layer withstands the voltage increase (Fig. 4.6).

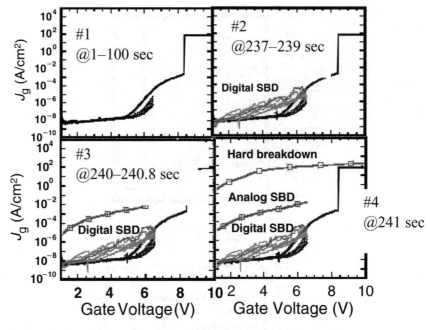

FIGURE 4.5: Breakdown procedure with time under substrate injection. It shows fast transition from soft breakdown to hard breakdown compared to that of gate injection.

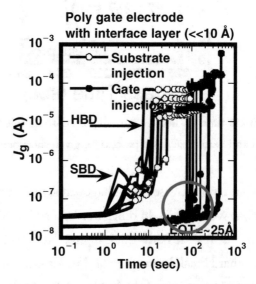

FIGURE 4.6: Worse soft breakdowns for substrate injection, but slow transition from soft to hard breakdown when it has very thin interface layer.

It has been suggested that the different physical and chemical nature of the interfacial layer and the bulk layer of HfO$_2$ may induce different breakdown behaviors [24, 26]. In general, single layer dielectric device like SiO$_2$ has shown an intrinsic and unimodal defect generation rate [29–32]. Therefore, thickness dependence of Weibull slope, voltage dependence of defect generation rate, thickness dependence of critical defect density, charge to breakdown statistics, and soft breakdown could be explained by percolation theory [29–32]. However, most of high-k dielectrics systems consist of interface layer and bulk high-k layer and apparently both layers have different physical and chemical properties. By separating or identifying different properties successfully, it is possible to support the theory that different natures of layers comprising high-k system influence device wear-out progresses. In this experiment, it is clearly observed that two-step breakdown occurs and both these breakdowns have quite different Weibull distribution behaviors with different workfunction gate electrode (Figs. 4.7 and 4.8).

It is observed that Weibull distribution shows not only slope β differences but also the difference of transition time between soft and hard breakdown. This

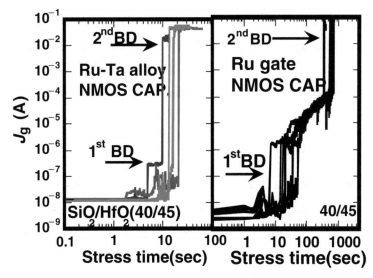

FIGURE 4.7: Two-step breakdowns for both Ru–Ta alloy and Ru gate due to bimodal defect generation. Higher barrier height by higher workfunction influences soft breakdown behaviors.

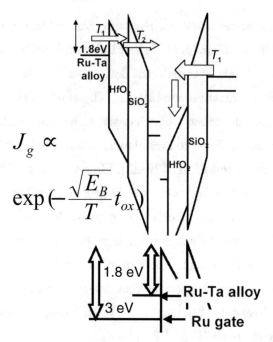

$$J_g \propto$$

$$\exp\left(-\frac{\sqrt{E_B}}{T}t_{ox}\right)$$

FIGURE 4.8: Band diagram for bimodal defect generation rate (T_1, T_2). Different charge fluence by polarity and barrier height.

two-step breakdown may be attributed to bimodal defect generation (T_1 and T_2) of interface and bulk high-k layer. The differences in the amount of leakage current of soft breakdown and transition time between soft and hard breakdown results from different amount of charge fluence due to the barrier height by gate workfunctions (Fig. 4.8).

4.4   CONCLUSION

The interface thickness is the one of key parameters to evaluate high-k gate stacks. It influences breakdown wear-out process as well as β. It is also worth emphasizing that low barrier height of high-k gate stacks increases soft breakdown, and breakdown lifetime of soft and hard breakdown. It has been reported that there is a universal relation between dielectric constant, barrier height, and electric breakdown strength [33]. As suggested in [33], different material structures with regard to dielectric constants might be the possible reason of the universal relation. In addition to this material structure, we have to consider different charge fluence, which directly influence dielectric breakdown. In other word, low barrier height of high-k dielectrics could be the reason why we observed such a universal relation. We strongly suggest that high trapping and weak breakdown strength may be attributed to low barrier height of high-k system.

REFERENCE

[1] P. K. Roy and I. C. Kizilyalli, "Stacked high-ε gate dielectric for gigascale integration of metal-oxide-semiconductor technologies," *Appl. Phys. Lett.*, Vol. 72, pp. 2835–2837, 1998. doi:10.1063/1.121473

[2] D. Park, Y.-C. King, Q. Lu, T.-J. King, C. Hu, A. Kalnitsky, S.-P. Tay, and C.-C. Cheng, "Transistor characteristics with Ta$_2$O$_5$ gate dielectric," *IEEE Electron Device Lett.*, Vol. 19, pp. 441–443, 1998. doi:10.1109/55.663533

[3] H. F. Luan, S. J. Lee, C. H. Lee, S. C. Song, Y. L. Mao, Y. L. Mao, Y. Senzaki, D. Roberts, and D. L. Kwong, "High quality Ta$_2$O$_5$ gate dielectrics with Tox,eq<10 Å," *IEDM Tech. Dig.*, pp. 141–144, 1999.

[4] A. Chatterjee, R. A. Chapman, K. Joyner, M. Otobe, S. Hattangady, M. Bevan, G. A. Brown, H. Yang, Q. He, D. Rodgers, S. J. Fang, R. Kraft, A. L. P. Rotondaro, M. Terry, K. Brennan, S.-W. Aur, J. C. Hu, H.-L. Tsai, P. Jones, G. Wilk, M. Aoki, M. Rodder, and I.-C. Chen, "CMOS metal replacement gate transistors using tantalum pentoxide gate insulator," *IEDM Tech. Dig.*, pp. 777–780, 1998.

[5] H.-S. Kim, D. C. Gilmer, S. A. Campbell, and D. L. Polla, "Leakage current and electrical breakdown in metal-organic chemical vapor deposited TiO$_2$ dielectrics on silicon substrates," *Appl. Phys. Lett.*, Vol. 69, pp. 3860–3862, 1996. doi:10.1063/1.117129

[6] B. H. Lee, Y. Jeon, K. Zawadzki, W.-J. Qi, and J. C. Lee, "Effects of interfacial layer growth on the electrical characteristics of thin titanium oxide films on silicon," *Appl. Phys. Lett.*, Vol. 74, pp. 3143–3145, 1999. doi:10.1063/1.124089

[7]   C. Hobbs, R. Hedge, B. Maiti, H. Tseng, D. Gilmer, P. Tobin, O. Adetutu, F. Huang, D. Weddington, R. Nagabushnam, D. O'Meara, K. Reid, L. La, L. Grove, and M. Rossow, "Sub-quarter micron CMOS process for TiN-gate MOSFETs with TiO_2 gate dielectric formed by titanium oxidation," *Symp. VLSI Tech. Dig.*, pp. 133–134, 1999.

[8]   K. J. Hubbard and D. G. Schlom, "Thermodynamic stability of binary oxides in contact with silicon," *J. Mat. Res.*, Vol. 11, pp. 2757–2776, 1996.

[9]   R. D. Shannon, "Dielectric polarizabilities of ions in oxides and fluorides," *J. Appl. Phys.*, Vol. 73, pp. 348–366, 1993. doi:10.1063/1.353856

[10] J. Robertson, "Band offsets of wide-band-gap oxides and implications for future electronic devices," *J. Vac. Sci. Tech. B*, Vol. 18, pp. 1785–1791, 2000. doi:10.1116/1.591472

[11] W.-J. Qi, R. Nieh, B. H. Lee, L. Kang, Y. Jeon, K. Onishi, T. Ngai, S. Banerjee, and J. C. Lee, "MOSCAP and MOSFET characteristics using ZrO_2 gate dielectric deposited directly on Si," *IEDM Tech. Dig.*, pp. 145–148, 1999.

[12] B. H. Lee, L. Kang, W.-J. Qi, R. Nieh, Y. Jeon, K. Onishi, and J. C. Lee, "Ultrathin hafnium oxide with low leakage and excellent reliability for alternative gate dielectric application," *IEDM Tech. Dig.*, pp. 133–136, 1999.

[13] W.-J. Qi, R. Nieh, B. H. Lee, K. Onishi, L. Kang, Y. Jeon, J. C. Lee, V. Kaushik, B.-Y. Neuyen, L. Prabhu, K. Eigenbeiser, and J. Finder, "Performance of MOSFETs with ultra thin ZrO_2 and Zr silicate gate dielecrics," *Symp. VLSI Tech. Dig.*, pp. 40–41, 2000.

[14] B. H. Lee, R. Choi, L. Kang, S. Gopalan, R. Nieh, K. Onishi, and J. C. Lee, "Characteristics of TaN gate MOSFET with ultra thin hafnium oxide," *IEDM Tech. Dig.*, pp. 39–42, 2000.

[15] L. Kang, Y. Jeon, K. Onishi, B. H. Lee, W.-J. Qi, R. Nieh, S. Gopalan, and J. C. Lee, "Single-layer thin HfO_2 gate dielectric with n+-polysilicon gate," *Symp. VLSI Tech. Dig.*, pp. 44–45, 2000.

[16] C. H. Lee, H. F. Luan, W. P. Bai, S. J. Lee, T. S. Jeon, Y. Senzaki, D. Roberts, and D. L. Kwong, "MOS characteristics of ultra thin rapid thermal

CVD ZrO$_2$ and Zr silicate gate dielectrics," *IEDM Tech. Dig.*, pp. 27–30, 2000.

[17]   M. Koyama, K. Suguro, M. Yoshiki, Y. Kamimuta, M. Koike, M. Ohse, C. Hongo, and A. Nishiyama, "Thermally stable ultra-thin nitrogen incorporated ZrO$_2$ gate dielectric prepared by low temperature oxidation of ZrN," *IEDM Tech. Dig.*, pp. 459–462, 2001.

[18]   R. Nieh, S. Krishnan, H.-J. Cho, C. S. Kang, S. Gopalan, K. Onishi, R. Choi, and J. C. Lee, "Comparison between ultra-thin ZrO$_2$ and ZrO$_x$N$_y$ gate dielectrics in TaN or poly-gated NMOSCAP and NMOSFET devices," *Symp. VLSI Tech. Dig.*, pp. 186–187, 2002.

[19]   M. T. Thomas, "Preparation and properties of sputtered hafnium and anodic HfO$_2$ films," *J. Electrochem. Soc.*, Vol. 117, pp. 396–403, 1970.

[20]   K. Kukli, J. Ihanus, M. Ritala, and M. Leskela, "Tailoring the dielectric properties of HfO$_2$-Ta$_2$O$_5$ nanolaminates," *Appl. Phys. Lett.*, Vol. 68, pp. 3737–3739, 1996. doi:10.1063/1.115990

[21]   R. Choi, K. Onishi, C. S. Kang, S. Gopalan, R. Nieh, Y. H. Kim, J. H. Han, S. Krishnan, H. J. Cho, A. Sharriar, and J. C. Lee, "Fabrication of high quality ultra-thin HfO$_2$ gate dielectric MOSFETs using deuterium anneal," *Tech. Dig. IEDM*, pp. 613–616, 2002.

[22]   K. Onishi, R. Choi, C. S. Kang, H. J. Cho, S. Gopalan, R. Nieh, S. Krishnan, and J. C. Lee, "Effects of high-temperature forming gas anneal on HfO$_2$ MOSFET performance," *Tech. Dig. Int. Symp. VLSI.*, pp. 22–23, 2002.

[23]   H. R. Huff, A. Hou, C. Lim, Y. Kim, J. Barnett, G. Bersuker, G. A. Brown, C. D. Young, P. M. Zeitzoff, J. Gutt, P. Lysaght, M. I. Gardner, and R. W. Murto, "Integration of high-*k* gate stacks into planar, scaled CMOS integrated circuits," in Conf. Nano and Giga Challenges in Microelectronics, 2002, pp. 1–18.

[24]   Y. H. Kim, K. Onishi, C. S. Kang, R. Choi, H.-J. Cho, R. Nieh, J. Han, S. Krishnan, A. Shahriar, and J. C. Lee, "Hard and soft-breakdown

characteristics of ultra-thin HfO$_2$ under dynamic and constant voltage stress," *Tech. Dig. IEDM*, p. 629, 2002.

[25] T. Kauerauf, R. Degraeve, E. Cartier, C. Soens, and G. Groesenenken, "Low Weibull slope of breakdown distributions in high-*k* layers," *IEEE Elec. Device Lett.*, Vol. 23, pp. 215–217, 2002. doi:10.1109/55.992843

[26] Y.-H. Kim, K. Onishi, C. S. Kang, H.-J. Cho, R. Choi, S. Krishnan, Md. S.r Akbar, and J. C. Lee, "Thickness dependence of Weibull slopes of HfO$_2$ gate dielectrics," *IEEE Elec. Device Lett.*, pp. 40–42, 2003.

[27] L. Kang, K. Onishi, Y. Jeon, B. Lee, C. Kang, W. Qi, R. Nieh, S. Gopalan, R. Choi, and J. C. Lee, "MOSFET devices with polysilicon on single-layer HfO$_2$ high-*k* dielectrics," *Tech. Dig. IEDM*, pp. 35–38, 2000.

[28] J. Lee, H. Zhong, Y.-S. Suh, G. Heuss, J. Gurganus, B. Chen, and V. Misra, "Tunable work function dual metal gate technology for bulk and non-bulk CMOS," *Tech. Dig. IEDM*, pp. 359–362, 2002.

[29] R. Degraeve, G. Groeseneken, R. Bellens, M. Depas, and H. E. Maes, "A consistent model for the thickness dependence of intrinsic breakdown in ultra-thin oxides," *Tech. Dig. IEDM*, pp. 863–866, 1995.

[30] M. Houssa, T. Nigam, P. W. Mertens, and M. M. Heyns " Model for the current-voltage characteristics of ultra thin gate oxide after soft breakdown" *J. App. Phys.*, Vol. 84, pp. 4351–4355, 1998. doi:10.1063/1.368654

[31] J. H. Stathis and D. J. DiMaria, " Reliability projection for ultra-thin oxides at low voltage," *Tech. Dig. IEDM*, pp. 167–170, 1998.

[32] J. H. Stathis, "Physical and predictive models of ultra thin oxide reliability in CMOS devices and circuits," IRPS, pp. 132–149, 2001.

[33] J. McPherson, J. Kim, A. Shanware, H. Mogul, and J. Rodriguez, "Proposed universal relationship between dielectric breakdown and dielectric constant," *Tech. Dig. IEDM*, pp. 633–636, 2002.

CHAPTER 5

Bimodal Defect Generation Rate by Low Barrier Height and its Impact on Reliability Characteristics

5.1 MOTIVATION

The origin of traps in high-k dielectrics, which play an important role in dielectric wear-out, however, still remains a question. It has been reported that Weibull slope of HfO_2 is smaller than that of SiO_2 with the same physical thickness [22, 23]. Possible explanations were suggested that this might be either smaller critical defect densities due to preexisting traps, extrinsic defects, or bimodal defect generations [22–25]. On the other hand, metal gate electrodes are expected to be needed in the near future. However, there has been a lack of the experimental evidences and study on the gate electrode effects on wear-out characteristics of high-k devices. In this chapter, we study the wear-out characteristics and Weibull slope behaviors of HfO_2 devices in terms of interface layer, high-k layer, and barrier height by gate electrode workfunction.

In this work, we present the effects of barrier height on the reliability of HfO_2 with dual metal gate technology in terms of Weibull slope, soft breakdown characteristics, defect generation rate, critical defect density, and charge-to-breakdown. It was found that the lower Weibull slope of high-k dielectrics (compared to SiO_2) is partially attributed to the lower barrier height of high-k dielectrics, which in turn results in larger current increase. Thus, defect generation rate increases and charge-to-breakdown decreases, while critical defect density remains constant. In addition, it has been found that there is distinct bimodal defect generation rate for high-k/SiO_2 stack. Two-step breakdown process was clearly observed; and Weibull slope of soft breakdown (first breakdown) shows lower β value compared to that of hard breakdown (second breakdown). Soft breakdown characteristics were dependent on the barrier heights. The bimodal defect generations are believed result from the breakdown in interface and bulk layer.

5.2   EXPERIMENTAL PROCEDURE

HfO_2 was deposited using reactive dc magnetron sputtering with O_2 modulation technique [26], followed by postdeposition annealing (PDA) at 500°C. Ru and Ru–Ta gate electrodes were deposited on top of HfO_2. W was deposited on the sample with Ru–Ta gate. The electrodes were patterned using lift-off process, followed by forming gas annealing at 400°C for 30 min. To keep the samples from possible metal diffusion into dielectrics, no high temperature processing was performed after gate electrode formation.

5.3   RESULTS AND DISCUSSION

Figure 5.1 shows the cross section of the device used in this study. Dual metal gate was achieved on the single wafer with the same gate dielectrics stacks and same type of substrate. Workfunctions of Ru and Ru–Ta alloy are shown in Fig. 5.2. Workfunction difference between Ru and Ru–Ta alloy is ∼1 eV [27]. For gate injection, higher gate leakage current was observed for Ru–Ta alloy gate because of the lower barrier height for electrons (Fig. 5.3), whereas for substrate injection, there is no significant change in leakage current at tunneling regime (Fig. 5.4). This is because gate leakage current (i.e. charge fluence) is strongly dependent on

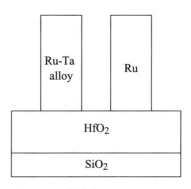

FIGURE 5.1: Dual metal gate was deposited on the single wafer with the same gate dielectric stacks.

conduction band offset for electron to surmount. That is why for gate injection, significant current change corresponding to conduction band offset was observed, whereas no significant change was observed for substrate injection.

Higher breakdown voltages were observed for Ru gate device under gate injection because of higher barrier height compared to that of Ru–Ta alloy gate (Fig. 5.5). It was suggested that breakdown in a wide range of high-k dielectrics follows closely a $(k)^{-1/2}$ dependence. A physical model was presented, which

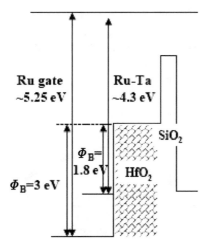

FIGURE 5.2: Workfunctions of dual metal gate and corresponding barrier heights.

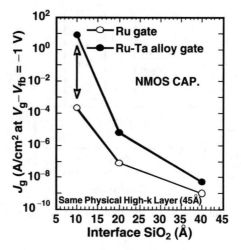

FIGURE 5.3: Higher leakage current for Ru–Ta alloy due to low barrier height.

suggested that the high local electric field (in high-*k* dielectrics) is very important in the breakdown process and drives the approximate $(k)^{-1/2}$ behavior. The high local field tends to distort/weaken polar bonds thereby making them very susceptible to breakage by standard Boltzmann processes [28]. It might be the good reason why we observed higher breakdown voltage for higher conduction band offset. That is, because of the universal relation between dielectric constant, conduction band

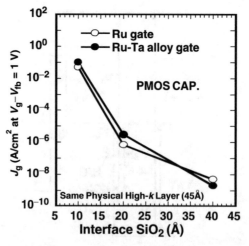

FIGURE 5.4: For substrate injection, there is no significant change in tunneling current.

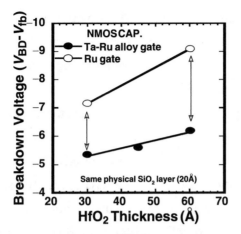

FIGURE 5.5: Breakdown voltages were increased for Ru gate due to higher barrier height.

offset, and breakdown strength, low barrier height might be more susceptible to breakdown. However, the physical structure model did not include the possibility of charge fluence by different conduction band offset because, at such a low voltage stress, it is worth empathizing that defect generation is attributed to charge fluence through a dielectric.

On the contrary, for substrate injection, similar breakdown voltages were observed for both Ru and Ru–Ta alloy gates (Fig. 5.6). This may result from no

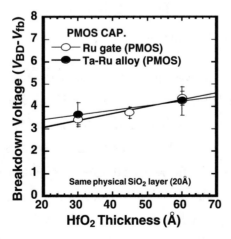

FIGURE 5.6: For substrate injection, no significant breakdown voltage change (same barrier height at substrate).

change in barrier height from a substrate side whatever gate electrode is when the device is subject to substrate injection. Therefore, asymmetric band structure plays an important role in triggering breakdown wear-out process in conjunction with different charge fluence by different barrier height.

The different physical and chemical nature of the interfacial layer and the bulk layer of HfO_2 may induce different breakdown behaviors. It can clearly be observed that two-step breakdown behavior occurs and that different Weibull distributions exist for different workfunction gate electrode (Figs. 5.7 and 5.8). This is the strong evidence of bimodal defect generation rate due to different nature of interface SiO_2 and bulk HfO_2. Unlike SiO_2, soft breakdown of HfO_2 may not be a weak percolative path between substrate and gate electrode, but it may be single layer breakdown among multiple layers.

Once one layer broke down, additional stress potential would be transferred to another layer, and it will increase the defect generation rate (N_{BD}) from initial N_{BD} (Fig. 5.9). Of course, critical defect density for each layer will not change whether stress voltage becomes high or not.

$$P_g = (\Delta J)/(J \Delta Q_{inj})$$
$$Q = \int J_g \, dt$$
$$Q_{BD} = N_{BD}/P_g$$

As explained earlier, defect generation rate, charge to breakdown, and critical defect density can be defined and extracted from SILC. Following previous SiO_2 study, electrons or holes tunneling through the gate oxide generate defects until a critical density is reached and the oxide breaks down. A percolation model can explain the critical defect density and the thickness dependence of the Weibull slope [29–32]. The nature of the electrical conduction through a breakdown spot will have a significant bearing on the degree to which the oxide breakdown affects device and circuit performance. The definition of oxide breakdown has generally been the first event marked by a discrete jump in leakage current. However, sometimes

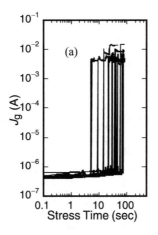

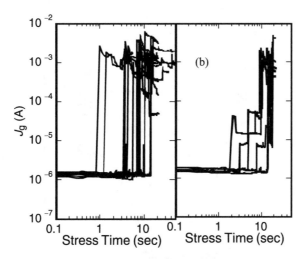

FIGURE 5.7: Soft breakdown becomes larger fraction of breakdown distributions for (a) Ru gate compared to that of (b) Ru–Ta alloy. High soft breakdown current results from higher barrier height between Ru gate and HfO_2.

the magnitude of this jump may not be great enough to completely destroy the functionality of a transistor, much less that of an entire circuit. It is often called "soft breakdown." In this study, we calculated breakdown parameters by separating soft breakdown and hard breakdown. It is believed that soft breakdown of high–k system would be neither weak percolation path of entire thickness nor localized

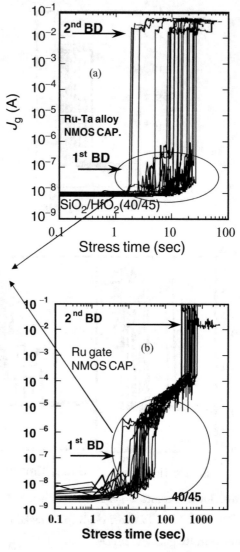

FIGURE 5.8: (a, b) Evolution of soft breakdown corresponding to barrier height. (Cont.)

conduction path if interface layer thickness is thick enough. It must be representing one layer breakdown of high-*k* dielectric structure. However, if interface layer is too thin, it would be hard to identify each layer breakdown.

In order to clarify different defect generation between two layers, we have grown different interface layer thicknesses ranging from 10 to 40 Å using RTP

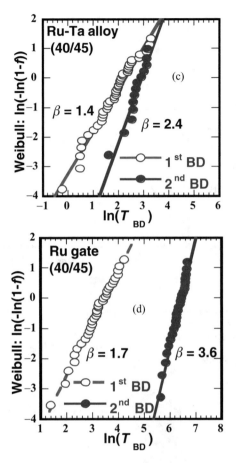

FIGURE 5.8: (Cont.) (c, d) Evolution of soft breakdown and hard breakdown and distinct Weibull distribution regarding barrier height.

system at 950°C. For given fixed interface layer, various HfO_2 thicknesses were deposited on top of intentional interface layer. Varying interface SiO_2 layers with the same high-k layer thickness shows strong thickness dependence of Weibull slope values β (Figs. 5.10–5.12), whereas varying high-k layers with the same interface layer exhibits weak function of thickness but clearly observed (Fig. 5.13). It is also observed that high barrier height increases Weibull slope value β mainly due to reduced charge fluence (Figs. 5.11 and 5.13). These results indicate that low Weibull slope values of high-k gate stack are due to not only low critical defect density by preexisting traps but also low barrier height of high-k dielectrics.

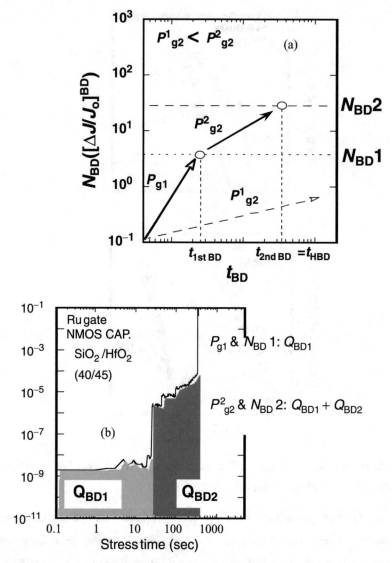

FIGURE 5.9: (a, b) Critical defect density and bimodal defect generations. Once the one layer broke down, additional stress voltage would be transferred to the other layer, and it will increase defect generation rate from P_{g2}^1 to P_{g2}^2, while critical defect density (N_{BD}) is constant.

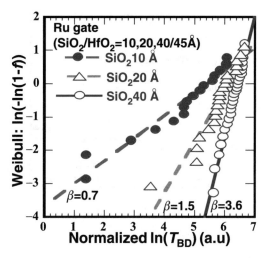

FIGURE 5.10: Normalized Weibull distribution of HfO_2 gate stack with varying SiO_2.

The concept of a critical defect density was quantitatively examined by Suñé *et al.* [33], who showed that it leads to the correct statistical behavior. Degraeve [29] suggested the percolation model in which breakdown is explained as the formation of a connecting path of defects, as a result of random defect generation throughout the insulating film. This explains the thickness dependence of N_{BD} for thinner

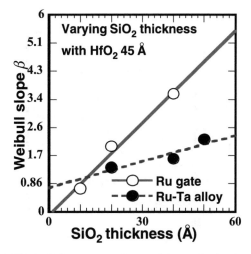

FIGURE 5.11: High Weibull slopes were observed due to higher barrier height of Ru gate compared to that of Ru–Ta alloy gate.

FIGURE 5.12: Weibull slopes show strong function of interface layer thickness, while weak function of high-k layer.

films. The percolation concept, and the origin of the thickness dependence of N_{BD}, is schematically illustrated by the computer simulation [34]. According to this model for single SiO_2 layer, breakdown can occur only when percolation path of traps is formed across the gate oxide, leading to a conducting path from the substrate to the gate. The probability of forming such a connecting path (percolation path) with randomly generated defects throughout the oxide bulk is computed as a

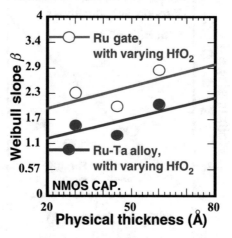

FIGURE 5.13: Weibull slope dependence of barrier height with varying HfO_2. Dual metal gate were deposited on the single wafer with the same gate dielectric stacks.

function of defect density and oxide thickness. However, because of nature of high-k system, bilayered structure requires more detailed and complicated analysis on wear-out process, in which bimodal defect generation rate may influence dielectric degradation. Fig. 5.14 shows that critical defect densities have no dependence of barrier height but strong function of interface SiO_2 layer. These results indicate that critical defect density is strongly dependent on each material property and thickness, rather than amount of charge fluence or stress voltage.

Lower defect generation rate (N_{BD}) with varying HfO_2 were observed for Ru gate compared to that of Ru–Ta alloy gate, whereas similar N_{BD} were observed for varying SiO_2 with both Ru and Ru–Ta alloy (Figs. 5.15 and 5.16).

For given critical defect density of each layer, amount of charge fluence significantly influences defect generation rate with respect to a barrier height. In other word, it is important to know critical defect density of each layer with function of thickness and different charge fluence must be taken into consideration because, in general, high-k dielectrics have low barrier height. It implies that soft and hard breakdowns depend strongly on the ratio of high-k layer to interface layer

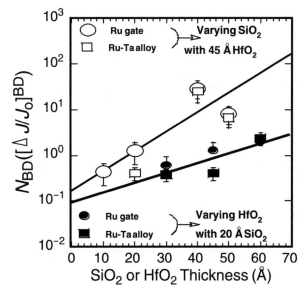

FIGURE 5.14: Critical defect density (N_{BD}) of varying SiO_2 and HfO_2 layers. N_{BD} is strong function of interface layer rather than HfO_2 layer.

FIGURE 5.15: Lower defect generation rates were observed for Ru gate compared to that of Ru–Ta alloy gate, as HfO_2 thickness changes. It is attributed to smaller charge fluence of Ru gate devices due to higher barrier height.

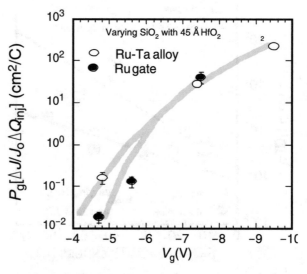

FIGURE 5.16: Similar defect generation rates were observed for either varying SiO_2 with 45 Å HfO_2 or varying HfO_2 with 20 Å SiO_2.

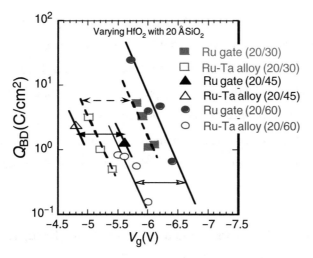

FIGURE 5.17: Charge-to-breakdown were extracted and compared to different barrier heights. Higher Q_{BD} is shown for Ru gate devices because it has low defect generation rate compared to that of Ru–Ta alloy.

thicknesses. With devices of varying HfO_2 thickness, higher charge-to-breakdown (Q_{BD}) was observed for Ru gate electrode due to higher barrier height and thus lower charge fluence (Figs. 5.17 and 5.18).

5.4   CONCLUSION

In this work, the effects of barrier height and the nature of the bilayer structure on the reliability of HfO_2 with dual metal gates was reported for the first time. The

$$P_{total} = P1 + P2$$
$$N_{BD} (\text{Å}) = N_{BD1} (\text{Å}) + N_{BD2} (\text{Å})$$

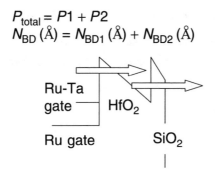

FIGURE 5.18: The effect of different nature of bi-layer structure and low barrier height of high-k dielectrics.

bimodal defect generation rate of HfO_2 was analyzed. It is quantitatively shown that large charge fluence caused by low barrier height could change Weibull slope β, soft breakdown and hard breakdown, defect generation rate, and charge-to-breakdown. The interface thickness is the one of key parameters to evaluate high-k gate stacks. It influences breakdown wearout process as well as Weibull slope β. It is also worth emphasizing that a low barrier height of high-k gate stacks can lead to lower Weibull slopes and shorter breakdown lifetimes of both soft and hard breakdowns.

REFERENCE

[1] T.-S. Chen, D. Hadad, V. Valu, V. Jiang, S.-H. Kuah, P. C. McIntyre, S. R. Summerfelt, J. M. Anthony, and J. C. Lee, "Ir-electroded BST thin film capacitors for 1 giga-bit DRAM application," in *IEEE International Electron Devices Meeting*, 1996, pp. 679–682.

[2] D. Barlage, R. Arghavani, G. Deway, M. Doczy, B. Doyle, J. Kavalieros, A. Murthy, B. Roberds, P. Stokley, and R. Chau, "High-frequency response of 100 nm integrated CMOS transistors with high-k gate dielectrics," in *IEEE International Electron Devices Meeting*, 2001, pp. 231–234.

[3] M. Balog, M. Schieber, M. Michman, and S. Patai, "Chemical vapor deposition and characterization of HfO_2 films from organo-hafnium compounds," *Thin Solid Films*, Vol. 41, pp. 247–259, 1997. doi:10.1016/0040-6090(77)90312-1

[4] B. Cheng, M. Cao, R. Rao, A. Inani, P. V. Voorde, W. M. Greene, J. M. C. Stork, Z. Yu, P. M. Zeitzoff, and J. C. S. Woo, "The impact of high-k gate dielectrics and metal gate electrodes on sub-100 nm MOSFET's," *IEEE Trans. Electron Devices*, Vol. 46, pp. 1537–1544, 1999. doi:10.1109/16.772508

[5] A. Kumar, T. H. Ning, M. V. Fischetti, and E. Gusev, "Hot-carrier charge trapping and reliability in high-k dielectrics," *Tech. Dig. VLSI Symp.*, pp. 152–153, 2002.

[6] M. Koyama, K. Suguro, M. Yoshiki, Y. Kamimuta, M. Koike, M. Ohse, C. Hongo, and A. Nishiyama, "Thermally stable ultra-thin nitrogen

incorporated ZrO_2 gate dielectric prepared by low temperature oxidation of ZrN," *Tech. Dig. IEDM,* pp. 459–462, 2001.

[7]   C. H. Lee, H. F. Luan, W. P. Bai, S. J. Lee, T. S. Jeon, Y. Senzaki, D. Roberts, and D. L. Kwong, "MOS characteristics of ultra thin rapid thermal CVD ZrO_2 and Zr silicate gate dielectrics," *Tech. Dig. IEDM,* pp. 27–30, 2000.

[8]   L. Kang, K. Onishi, Y. Jeon, B. Lee, C. Kang, W. Qi, R. Nieh, S. Gopalan, R. Choi, and J. C. Lee, "MOSFET devices with polysilicon on single-layer HfO_2 high-*k* dielectrics," *Tech. Dig. IEDM,* pp. 35–38, 2000.

[9]   B. H. Lee, R. Choi, L. G. Kang, S. Gopalan, R. Nieh, K. Onishi, Y. J. Jeon, W. J. Qi, C. S. Kang, and J. C. Lee, "Characteristics of TaN gate MOSFET with ultrathin hafnium oxide (8 Å–12 Å)," *Tech. Dig. IEDM,* p. 39, 2000.

[10] J. Robertson, "Band offsets of wide-band-gap oxides and implications for future electronic devices," *J. Vacuum Sci. Technol. B,* Vol. 18, pp. 1785–1791, 2000. doi:10.1116/1.591472

[11] R. Choi, K. Onishi, C. S. Kang, S. Gopalan, R. Nieh, Y. H. Kim, J. H. Han, S. Krishnan, H. J. Cho, A. Sharriar, and J. C. Lee, "Fabrication of high quality ultra-thin HfO_2 gate dielectric MOSFETs using deuterium anneal," *Tech. Dig. IEDM,* pp. 613–616, 2002.

[12] K. Onishi, R. Choi, C. S. Kang, H. J. Cho, S. Gopalan, R. Nieh, S. Krishnan, and J. C. Lee, "Effects of high-temperature forming gas anneal on HfO_2 MOSFET performance," *Tech. Dig. Int. Symp. VLSI.,* pp. 22–23, 2002.

[13] H. R. Huff, A. Hou, C. Lim, Y. Kim, J. Barnett, G. Bersuker, G. A. Brown, C. D. Young, P. M. Zeitzoff, J. Gutt, P. Lysaght, M. I. Gardner, and R. W. Murto, "Integration of high-*k* gate stacks into planar, scaled CMOS integrated circuits," in *Conf. Nano and Giga Challenges in Microelectronics,* pp. 1–18, 2002.

[14] W.-D. Kim, J.-H. Joo, Y.-K. Jeong, S.-J. Won, S.-Y. Park, S.-C. Lee, C.-Y. Yoo, S.-T. Kim, and J.-T. Moon, "Development of CVD-Ru/Ta,O,/CVD-Ru capacitor with concave structure for multigigabit-scale DRAM generation," *Tech. Dig. IEDM,* pp. 12.1.1–12.1.4, 2002.

[15] H. Seidl, M. Gutsche, U. Schroeder, A. Birner, T. Hecht, S. Jakschik, J. Luetzen, M. Kerber, S. Kudelka, T. Popp, A. Orth, H. Reisinger, A. Saenger, K. Schupke, and B. Sell, "A fully integrated Al_2O_3 trench capacitor DRAM for sub-100 nm technology, " *Tech. Dig. IEDM*, pp. 839–842, 2002.

[16] C. B. Oh, H. S. Kang, H. J. Ryu, M. H. Oh, H. S. Jung, Y. S. Kim, J. H. He, N. I. Lee, K. H. Cho, D. H. Lee, T. H. Yang, I. S. Cho, H. K. Kang, Y. W. Kim, and K. P. Suh, "Manufacturable embedded CMOS 6T-SRAM technology with high-k gate dielectric device for system-on-chip applications," *Tech. Dig. IEDM*, pp. 423–426, 2002.

[17] Y. Matsui, M. Hiratani, I. Asano, and S. Kimura, "Niobia-stabilized tantalum pentoxide (NST)—Novel high-k dielectrics for low-temperature process of MIM capacitors," *Tech. Dig. IEDM*, pp. 225–228, 2002.

[18] L. Manchanda, M. L. Green, R. B. van Dover, M. D. Morris, A. Kerber, Y. Hu, J.-P. Han, P. J. Silverman, T. W. Sorsch, G. Weber, V. Donnelly, K. Pelhos, F. Klemens, N. A. Ciampa, A. Kornblit, Y. O. Kim, J. E. Bower, D. Barr, E. Ferry, D. Jacobson, J. Eng, B. Busch, and H. Schulte, "Si-doped aluminates for high temperature metal-gate CMOS: Zr-Al-Si-O, a novel gate dielectric for low power applications," *Tech. Dig. IEDM*, pp. 23–26, 2000.

[19] T. Sugizald, M. Kobayashi, M. Ishidao, H. Minakata, M. Yamaguchi, Y. Tamura, Y. Sugiyarna, T. Nakanishi, and H. Tanaka, "Novel multi-bit sonos type flash memory using a high-k charge trapping layer," *Tech. Dig. Int. Symp. VLSI.*, pp. 27–28, 2003.

[20] Y. L. Tu, H. L. Lin, L. L. Chao, D. Wu, C. S. Tsai, C. Wang, C. F. Huang, C. H. Lin, and J. Sun, "Characterization and comparison of high-k metal-insulator-metal (mim) capacitors in 0.13 um Cu BEOL for mixed-mode and rf applications," *Tech. Dig. Int. Symp. VLSI.*, pp. 79–80, 2003.

[21] J. J. Lee, X. Wang, W. Bai, N. Lu, J. Lni, and D. L. Kwong, "Theoretical and experimental investigation of si nanocrystal memory device with HfO_2 high-k tunneling dielectric," *Tech. Dig. Int. Symp. VLSI.*, pp. 33–34, 2003.

[22] Y. H. Kim, K. Onishi, C. S. Kang, R. Choi, H.-J. Cho, R. Nieh, J. Han, S. Krishnan, A. Shahriar, and J. C. Lee, "Hard and soft-breakdown characteristics of ultra-thin HfO$_2$ under dynamic and constant voltage stress," *Tech. Dig. IEDM*, pp. 629–632, 2002.

[23] T. Kauerauf, R. Degraeve, E. Cartier, C. Soens, and G. Groesenenken, "Low Weibull slope of breakdown distributions in high-*k* layers," *IEEE Electron Device Lett.*, Vol. 23, pp. 215–217, 2002. doi:10.1109/55.992843

[24] Y.-H. Kim, K. Onishi, C. S. Kang, H.-J. Choi, R. Nieh, S. Gopalan, R. Choi, J. Han, S. Krishnan, and J. C. Lee, "Area dependence of TDDB characteristics for HfO$_2$ gate dielectrics," *IEEE Electron Device Lett.*, Vol. 23, pp. 594–596, 2002. doi:10.1109/LED.2002.1004214

[25] Y.-H. Kim, K. Onishi, C. S. Kang, H.-J. Cho, R. Choi, S. Krishnan, Md. S. Akbar, and J. C. Lee, "Thickness dependence of Weibull slopes of HfO$_2$ gate dielectrics," *IEEE Electron Device Lett.*, pp. 40–42, 2003.

[26] L. Kang, K. Onishi, Y. Jeon, B. Lee, C. Kang, W. Qi, R. Nieh, S. Gopalan, R. Choi, and J. C. Lee, "MOSFET devices with polysilicon on single-layer HfO$_2$ high-*k* dielectrics," *Tech. Dig. IEDM,* pp. 35–38, 2000.

[27] J. Lee, H. Zhong, Y.-S. Suh, G. Heuss, J. Gurganus, B. Chen, and V. Misra, "Tunable work function dual metal gate technology for bulk and non-bulk CMOS," *Tech. Dig. IEDM*, pp. 359–362, 2002.

[28] J. McPherson, J. Kim, A. Shanware, H. Mogul, and J. Rodriguez, "Proposed universal relationship between dielectric breakdown and dielectric constant," *Tech. Dig. IEDM*, pp. 634–637, 2002.

[29] R. Degraeve, G. Groeseneken, R. Bellens, M. Depas, and H. E. Maes, "A consistent model for the thickness dependence of intrinsic breakdown in ultra-thin oxides," *Tech. Dig. IEDM*, pp. 863–866, 1995.

[30] M. Houssa, T. Nigam, P. W. Mertens, and M. M. Heyns "Model for the current-voltage characteristics of ultra thin gate oxide after soft breakdown," *J. App. Phys.*, Vol. 84, pp. 4351–4355, 1998. doi:10.1063/1.368654

[31] J. H. Stathis and D. J. DiMaria, "Reliability projection for ultra-thin oxides at low voltage," *Tech. Dig. IEDM*, pp. 167–170, 1998.

[32] J. H. Stathis, "Physical and predictive models of ultra thin oxide reliability in CMOS devices and circuits," *IRPS proceeding of IEEE international reliability physics symposium*, pp. 132–149, 2001.

[33] J. Suñé, I. Placencia, N. Barniol, E. Farrés, F. Martín, and X. Aymerich, "On the breakdown statistics of very thin SiO_2 films," *Thin Solid Films*, Vol. 185, pp. 347–362, 1990. doi:10.1016/0040-6090(90)90098-X

[34] J. H. Stathis, "Percolation models for gate oxide breakdown," *J. Appl. Phys.*, Vol. 86, pp. 5757–5766, 1999. doi:10.1063/1.371590

The Authors

YOUNG-HEE KIM

Young-Hee Kim was born in Yang-Pyung, Korea, on January 24, 1972, as a son of Yong-Kae Kim and Jong-Rae Lee. He graduated from Sajic High School, Pusan, Korea, and joined Korean Air Force serving as military policeman. In 1995, he got admission in Kyung-Hee University, Suwon, Korea, and in 1999 he received the B.S. degree in electrical and computer engineering. In August 1999, he started his Masters and Ph.D. study in the department of electrical and computer engineering at The University of Texas at Austin. During the graduate work, he contributed to a number of technical publications (34 coauthored papers, in 14 of which he was the first author). In April 2004, he joined IBM T.J. Watson Research Center as a research staff member and has been carrying out exploratory devices integration.

JACK C. LEE

Jack C. Lee received the B.S. and M.S. degrees in electrical engineering from University of California, Los Angeles, in 1980 and 1981, respectively; and the Ph.D. degree in electrical engineering from University of California, Berkeley, in 1988.

He is a Professor of the Electrical and Computer Engineering Department and holds the Cullen Trust For Higher Education Endowed Professorship in Engineering at The University of Texas at Austin. From 1981 to 1984, he was a Member of Technical Staff at the TRW Microelectronics Center, CA, in the High-Speed Bipolar Device Program. He has worked on bipolar circuit design, fabrication, and testing. In 1988, he joined the faculty of The University of Texas at Austin. His current research interests include thin dielectric breakdown and reliability, high-k gate dielectrics and gate electrode, high-k thin films for semiconductor memory applications, electronic materials, and semiconductor device fabrication processes, characterization, and modeling. He has published over 300 journal publications and conference proceedings. He has received several patents. Dr. Lee has been awarded two Best Paper Awards and numerous Teaching/Research Awards. Dr. Lee is a Fellow of IEEE.

Printed in the United States
by Baker & Taylor Publisher Services